Designer's Guide to Creating
Charts & Diagrams

Designer's Guide to Creating
Charts & Diagrams
by Nigel Holmes

WATSON-GUPTILL PUBLICATIONS/NEW YORK

Paperback Edition
First Printing 1991

Copyright © 1984 by Watson-Guptill Publications

First published in 1984 by Watson-Guptill Publications,
a division of BPI Communications, Inc.,
1515 Broadway, New York, N.Y. 10036

Library of Congress Catalog Card Number: 84-3609
ISBN 0-8230-1315-4
ISBN 0-8230-1338-3 (pbk.)

Manufactured in U.S.A.

1 2 3 4 5 6 / 96 95 94 93 92 91

Edited by Stephen A. Kliment and Susan Davis
Designed by Bob Fillie
Graphic production by Hector Campbell
Set in 11 point Times Roman

Cover illustration:
Plates created on a (CAS) 2000 computer
by Collier Graphic Services

Cover design by Nigel Holmes

To Rowland

Contents

Acknowledgments

Many people have helped me with the writing and production of this book. At Watson-Guptill I especially thank Stephen Kliment for his planning, patience, and prodding into action. Many thanks also to Susan Davis who edited the book and along the way kept my American from slipping into English. Bob Fillie, the designer, and Hector Campbell of production waited a long time before their parts could be played. They never grumbled, at least not to me!

Nino Telak, the sort of assistant all chartmakers would kill for, helped me with the pieces drawn specially for this project; his mechanical assistance is also vital to the production of the weekly charts drawn for *Time*, many of which are reproduced here too. Also from *Time*, special mention must be made regarding Noel McCoy, Sara Noble, and Deborah Wells. Without the hard work and experience of these three researchers, none of the *Time* charts in this book could have existed. Often overlooked and seldom thanked, researchers provide the material with which to create. I am lucky and grateful to have worked with such a good team. Nick Hollis did the major part of the typing; Wendy Hanamura and Dorothy Chapman completed it.

At Collier Graphic Services, Michael Lissauer, Meline Follert, Ernie Miglaccio, Fred Kaselow, and Zane Tankel have pushed to new limits the technical side of producing the printed images and have generously allowed me to interfere in that process.

In England I would like to thank Brian Haynes, Peter Sullivan, and David Driver; in America, Ray Cave of *Time* magazine as well as Walter Bernard and Marshall Loeb. Although they were not directly involved with this book, they have, through their actions, advice and encouragement over a period of years shaped my thinking on the subject and more importantly have given me the opportunity to put the ideas into print.

To all those for whom I was either late or absent professionally or socially over the last two years, my apologies. This was the reason.

Preface

WE ARE DELUGED by a daily surge of figures: stock market figures, budget figures, deficit figures, population trends, increasing traffic trends, falling currency rates. In America every year:

The average person eats 81½ hot dogs and drinks 40 gallons of soda.

1.6 billion cigarettes are smoked.

There are 42 million pet cats.

The U.S. government is spending money at the rate of $23,000 a second.

How do we cope with them? Too many of us are scared by figures. We do not understand them. We would rather not look at them. We would like someone to explain them to us. Except to the trained few, figures are so anonymous, so flat, so obscure, and yet at the same time so threatening, as though they hide some secret that, if only we could see it, would reveal the horrible or wonderful truth of their subject.

The chartmaker can unlock the secrets of the figures. He or she can make their meaning visible, literally.

A simple chart is no more than a set of statistics made visible. It shows what has happened in the past and what might happen in the future.

But it can do more as well. It can engage the viewer by capturing his or her imagination. It can interpret the figures as well as present them. In fact, to simply parade the numbers as a set of bars or a rising and falling line does only half the job. It gives no clue as to the subject being dealt with. In certain contexts it might be perfectly proper to display the figures without any other visual help, but soon the charts will all look the same and therefore fail to be helpful, losing the interest of the consumer in much the same way as the bald figures themselves did before.

The purpose then of this book is to show:

That the role of the chartmaker can be greater than he or she ever thought.

That the job need not be boring, plodding plotting, but instead an enlightened experience for himself and those who use the work.

That those in a position to commission graphs, charts, and diagrams might expect more from their artists and might even be surprised by the results. They might laugh at the results. And thereby remember the salient facts.

That those about to hire an artist to work in this field might interview a mind first and a craftsman with a specific skill later.

That a writer might understand he or she is not losing precious lines of delicious prose to some unnecessary graphic, but instead be pleased that the attractiveness of the illustration will lead the viewer into reading the words. In the process he or she will be free to write unhindered by the statistics which are necessary for the argument but which get in the way of the article's flow.

The way some of this can be achieved is by joining together mathematics and illustration while keeping the statistics—the very reason for the existence of the chart—uppermost in the mind. The skillful chartmaker should attempt to find a suitable visual vehicle into which the facts can be placed. At the least this would be a background or frame for an otherwise perfectly conventional chart or graph, and at best it would be a marriage of statistics and illustrations, where the figures are present and the form they are displayed in reveal their subject.

The artist undertaking this task of welding together fact and picture should be prepared with arguments and reasons for so doing. He or she will not have an easy ride. He will be thought of as trivializing the nature of the figures he is portraying. He will be challenged on the correctness of the image chosen to accompany the figures; he will be questioned about why such an image is needed at all. His charts must speak for themselves of course, but he will also have to back them up with reasons at first, until he has won the confidence of client, editor, or reader. And having won that trust, he must be equally prepared to argue against the use of pictorial elements, for there are times when it will be more appropriate to do without them.

In short, the chartmaker should take charge of the situation as best he or she can under the particular circumstances, take the job seriously, and fully realize the potential of it.

Are you a magazine editor hoping to inform your readers of the history and present state of unemployment?

Are you a newspaper reporter working on a profile of crime in your city?

Are you a corporate V.P. in charge of your company's annual report that needs to show the breakdown of income and spending?

Do you work in the State Budget Office and have to prepare a large chart to show at a press conference and later at a meeting of interested citizen groups?

As an architect or designer, do you have to show your client the details of the costs of your plan for them? Or do you have to present a step-by-step guide to show how each segment of the construction of their office block will be dovetailed so that the night workers are on-site at the right time? Or must you show why the offices are so configured as to facilitate the flow of work required by the employers?

Do you publish a newsletter on oil and gas? Perhaps you need to show what steps increase the price of crude oil from the ground in Saudi Arabia through its journey to the gas you buy for your car?

Are you a management consultant who must demonstrate a more efficient working method for a company?

Do you need to show trends in infant mortality for a UNICEF exhibition traveling to junior high schools around the country? Or on an around-the-world trip, including places where no English is spoken? Or to university-level audiences?

Do you need a raise and feel that your situation would be better understood if presented graphically to your boss?

Are you preparing coverage of the elections for a local cable TV network?

Are you the President of the United States about to face the nation with unpopular but necessary budget proposals?

If you answered "yes" to any of these questions, this book will help you.

Chapter One
The Chart Starts Here

1	2	3	4	5	6	7	8	9	10
0	0	0	0	0	0	0	0	0	0
0	.	.	.	.	.	.	.	.	.
	0	.	.	.	.	.	.	.	.
		0	.	.	.	.	.	.	.
			0	.	.	.	.	.	.
				0	.	.	.	.	.
					0	.	.	.	.
						0	.	.	.
							0	.	.
								0	.
									0

Leonardo da Vinci plotted the fall of an object through space and time. His habit of writing everything from the left (top) is translated into modern type the more conventional way (bottom).

NO ONE THOUGHT much about charts before Roger Bacon came along. True, the first calendar of 365 days was invented by the Egyptians sometime between 2800 and 2600 B.C., Aristotle applied systematic thinking to science in 350 B.C., the Chinese invented paper in 100 A.D. and abacus calculators around 570, the Japanese printed from blocks with raised surfaces in the 8th century, and without these steps charts could never have been drawn. But only on Bacon's urging that mathematics be learned at universities did a means by which to study measurements begin. The work of this English scientist, who lived from 1214 to 1294, provided the impetus and inspiration that led, ultimately, to the science of visualizing abstract statistics—or charts.

Whereas astronomy had its own practical traditions of observation and measurement, there was no procedure for taking measurements connected with mechanics and physics, and thus those sciences developed along a theoretical path only. Important work in the early part of the 14th century was done in Oxford by Thomas Bradwardine, who used algebraic letters and symbols and laid the foundation for the next natural step—representing one function and its relation to another function by the use of graphs.

At about the same time in Paris, Albert of Saxony, Marsilius of Inghen, and, most importantly, Nicole d' Oresme, Bishop of Lisieux (1325–1382) discovered the logical equivalence between the organization of figures into groups and the graphing of those figures. Oresme himself produced the earliest graphs.

In 1465 the printing press first appeared in Italy. Although there is no direct evidence that Leonardo da Vinci (1452–1519) actually read the largely theoretical works of these early mathematicians from Oxford and Paris, the influence and traditions of their theories must have had a profound effect on him. In his notebooks Leonardo made one of the earliest versions of graphic representation. It plots a falling body against a time frame.

Since Leonardo wrote everything reversed from left to right I have set his graph the correct way round in modern type. Attached to the original is a note that reads, "Here is shown how such a proportion as one quantity of time and another will have a quantity of movement with the other, and a quantity of speed with the other."

DESCARTES' CONTRIBUTION

Later René Descartes (1596–1650) began to develop a systematic basis for this kind of thinking. In a 106-page footnote to his book *Discourse on the Method of Rightly Conducting the Reason and Seeking Truth in the Sciences* (published in 1637, 18 years after his original inspirational dream on November 10, 1619), Descartes outlined his "analytic geometry." This mere addendum to his philosophical theories was eventually to become the work that established him as more important in the field of science and mathematics than in that of philosophy. Analytic geometry had as its basis the idea that two numbers can describe the position of a point on a surface. He used two lines with scales marked on them called "coordinates"; the numbers were one of the points on the scale along each line. One of the lines was a distance measured horizontally, the other a distance measured vertically. Descartes' concept of a graph, using a grid of criss-crossing straight lines to describe a surface, was the forerunner of today's graph paper. His system of intersecting lines or coordinates is known as the "Cartesian grid." The original horizontal line is the x axis (or abscissa) and is the base line of the graph. The vertical line is the y axis (or ordinate).

Descartes' approach to geometry gave mathematicians a new way to look at equations, and is the basis of plotting statistics that we use today in charts and graphs.

PLAYFAIR LED THE WAY

John Playfair (1748–1819), a Scottish mathematician and a professor of maths (1785–1805) and natural philosophy (1805–1819) at the University of Edinburgh, noted in his *Progress of Natural Philosophy* that "it is seldom the destination of nature that a new discovery should be begun and perfected by the same individual and, in these attempts, though Descartes did not entirely fail, he cannot be considered as having been successful." Sir Isaac Newton and Gottfried Liebnitz separately furthered Descartes' theories of coordinates with reference to mathematical questions.

At about the same time as John Playfair was writing his theories his younger brother William was inventing the application of charts and graphs to financial statistics. Because he was determined to become an author, he dismissed his early training and associations with those involved in engineering and other practical sciences and thus did not achieve the eminence he deserved. What is more, he was extremely opinionated and held controversial ideas about government and reform, which he pressed on others. He thus fell into the gap between writer and scientist and was recognized as neither: The scientists of his day did not recognize his inventions, nor the literary set his writing. He is almost unknown today.

William Playfair was born near Dundee, Scotland, in 1759 and was apprenticed as a teenager to Andrew Meikle, the inventor of the threshing machine. Later he became highly skilled at technical drawing, developed while working with Boulton & Watt, the engineering firm, in Birmingham, England. This technical background obviously contributed to his discovery of graphic methods of charting.

When he was 25 in 1786, he published the *Commercial & Political Atlas.** It contains the first types of chart he perfected, the fever or line graph and the bar chart. There were forty-four charts in the "atlas," about half of them hand-colored copper-plate engravings representing trade between England and other countries. Playfair also included tables of the figures he was charting—"the mode here adopted for conveying information is accurate in its principle though in the execution

*In two later editions, in 1787 and 1801, the information in the original was updated.

it may be liable sometimes to error. I have however spared no pains to avoid mistakes and have added printed tables."

Playfair explains his method: "If the money received by a man in trade were all in guineas, and every evening he made a single pile of all the guineas received during the day, each pile would represent a day, and its height proportioned to the receipts of that day; so that by this plain operation, time, proportion and amount would all be physically combined. 'Lineal Arithmetic' then, it may be averred, is nothing more than those piles of guineas represented on paper, and on a small scale, in which an inch perhaps represents the thickness of five million guineas. As in geography it does the breadth of a river, or any other extent of the country. By this method as much information may be obtained in five minutes as would require whole days to imprint on the memory . . . by a table of figures."

He further points out that "The advantage proposed by those charts, is not that of giving a more accurate statement than by figures, but it is to give a more simple and permanent idea of the gradual progress and comparative amounts, at different periods, by presenting to the eye a figure, the proportions of which correspond with the amount of the sums intended to be expressed."

Playfair generally accompanied his charts with lengthy descriptions to underline the points being made. They were often used as visual evidence in a case he was championing. For instance, he wrote to the two Houses of Parliament about "our agricultural distresses, their causes and remedies, accompanied with tables and copper-plate charts, shewing and comparing the price of wheat, bread and labour, from 1565 to 1821" (published in London in 1821).

He charted the effect that wars had on Britain's national debts, showing very clearly by means of a fever chart, which he shaded in from the plotted line to the base, that the debt leapt up during the war periods. All Playfair's work was most carefully and methodically annotated, which today is not so necessary as it was when his charts first appeared before a public unaccustomed to reading anything like them. The national debt chart, for instance, has the note, "The Divisions at the Bottom are Years and those on the Right hand, Money." Every year from 1688 to 1784 is marked with a neat tick and every million pounds marked on the vertical scale from 0 to 300. And he added in an introduction to *An Inquiry into the Causes of the Decline and Fall of Powerful & Wealthy Nations* (1805), "the reader will find 5 minutes attention to the principle on which the charts are constructed a saving of much labour and time; but without that trifling attention, he may as well look at a blank sheet of paper as at one of the charts."

A French edition of the atlas published in 1789 was a great success. It is interesting to note that it was the French 400 years earlier who had taken some of the first steps toward graphic representations and who now were developing Playfair's inventions. For it was in France that much of the next century's graphic activity took place. Perhaps because Playfair himself underestimated the importance of his graphic work in favor of his writing, he did not further his own inventions (though there is evidence he was pleased by the reception they had in France). This is a pity since he seems to have been eminently qualified to be the forerunner of today's graphic journalist, coupling writing and drafting skills with a keen mind, open to seeing how groups of different statistics, when combined, could make a very telling point. He moved to France in 1787, but his political views and writings, especially attacks on the Constitution in 1789, led to his having to leave in 1792.

What of the people who saw his charts? His titling of the pieces with a large, bold, and often decoratively treated inscriptions is perhaps his

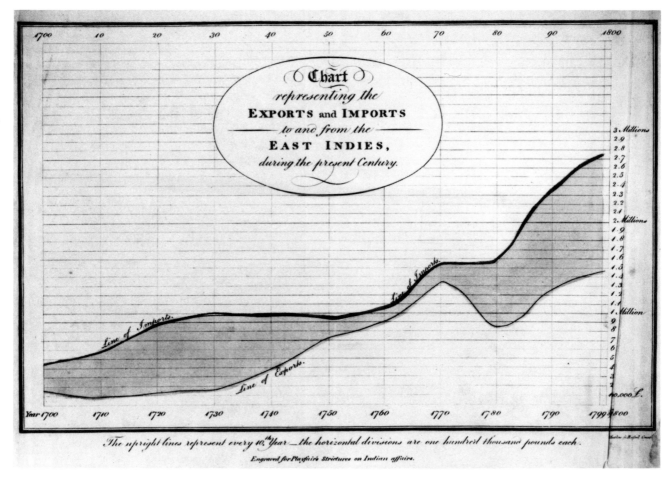

The upright lines represent every 10th Year — the horizontal divisions are one hundred thousand pounds each.

Engraved for Playfair's Strictures on Indian affairs.

The first use of fever charts was by William Playfair at the end of the 18th century. (Above) This chart represents the exports and imports to and from the East Indies between 1700 and 1799. Hand-colored, it is accompanied by an explanation of how to read the information and a warning that these are generalized graph lines that can never be as specific as a table of figures. From An Inquiry into the Causes of the Decline and Fall of Powerful and Wealthy Nations.

(Right) Playfair's chart showing English imports and exports to and from all North America was first published in 1785 in his Commercial and Political Atlas and showed how the balance of trade changed from being "in favour of England" at the left, to "against England" for a period during 1774–1775, then back to the original situation for the rest of the timespan. This uncolored version is from the second edition of the book, published in 1787. With only a slight sophistication in the use of type, charts are still drawn like this today. Playfair's original ideas have remained the basis for all modern chartmaking.

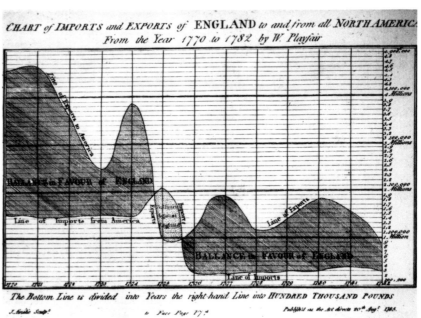

The Bottom Line is divided into Years the right-hand Line into HUNDRED THOUSAND POUNDS

15

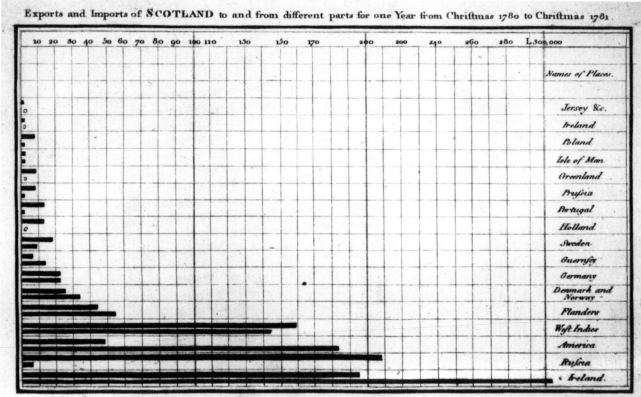

Exports and Imports of SCOTLAND to and from different parts for one Year from Christmas 1780 to Christmas 1781.

The Upright divisions are Ten Thousand Pounds each . The Black Lines are Exports the Ribbed lines Imports.

A bar chart showing Scotland's imports and exports from the Commercial and Political Atlas *(second edition, 1787). The bar is the only form of statistical presentation that Playfair used but did not invent. Here horizontal bars are grouped in twos, the upper "ribbed" (cross-hatched) bars representing imports and the lower black lines representing exports.*

In 1805 Playfair published the first pie chart, in the Statistical Account of the United States of America. *It is delicately colored by hand, in washes of light blue, green, and pink and shows proportional divisions of the United States at that time. The principle of the pie chart, which he called a "divided circle," is the same today as it was when Playfair invented it.*

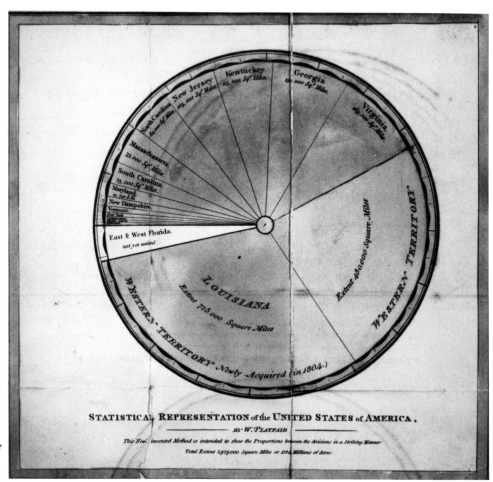

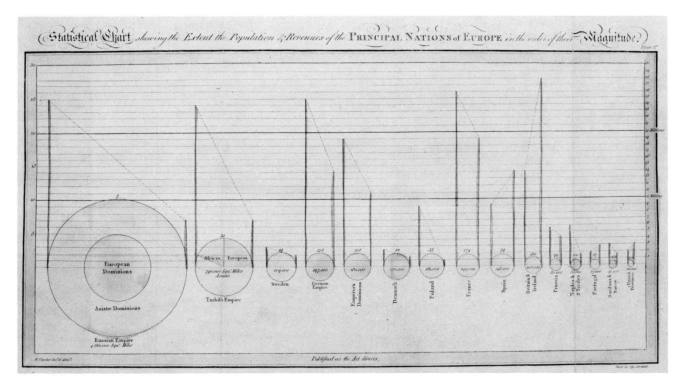

(Above) In 1801 Playfair produced the Statistical Breviary, *in which this chart appeared. His name for it was "circle graph." There is an enormous amount of data contained in this piece. The size of each circle is in proportion to the area of the nation it represents, with the uprights standing for population on the left and revenues on the right. Playfair mistakenly draws a conclusion from the slope of the line joining these two uprights. This slope of course depends as much on the width of the circle as on the height of the uprights to produce its angle.*

(Below) Also from the Statistical Breviary *is this diagram of European cities ranked in order of their population. It is interesting to note that in 1805 London's population was 2,300,000 compared with Moscow with 250,000. Present-day figures are London, 6,700,000, and Moscow, 8,200,000. New York City went from 50,000 to 7,000,000 in the same time.*

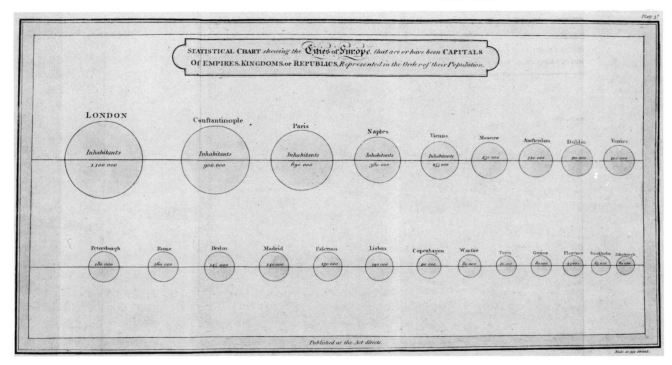

Fortune *was one of the first mass-circulation magazines to use illustrated charts. This divided bar chart from 1939 looks as fresh today as it must have then.*

version of the man with the red flag who walked in front of the first motor cars, warning the unsuspecting pedestrians of what was to come. We take the form for granted today and look for the story; in the 18th century the form was new. It has changed very little since then, but now we recognize it and do not need the red flag.

With the publication in 1801 of his *Statistical Breviary* Playfair introduced the circle chart, a form that is hard to use without distortion but that is nevertheless interesting and worthy of note. The size of each circle is proportionate to the area of a country. Two lines rise from the edges of the circle. That on the left measures the population of the country, on the right the revenue. Playfair makes the mistake of joining up the two with a diagonal line and suggests that this represents the tax burden. The slope of course depends not only on the height of the two bars that it joins, but also on the diameter of the circle.

1805 saw the publication of another of Playfair's chart inventions: the "divided circle." Nowadays we would call it a "pie chart." He translated into English "The Statistical Account of the United States of America" by the French statistician D. F. Donnant and added his own illustration, a chart "representing the proportional extent of the different states of the Eastern Country and the newly acquired territory of Louisiana." The pie chart was inscribed: "This newly invented method is intended to shew the proportions in a striking manner." He sent the book to Thomas Jefferson and asked that the President use the blank pages which he had intentionally bound into the volume to add his comments and alterations to any of the facts that were out of date or wrong. It was filled with statistics: In 1803, for instance, an individual could defray his year's taxes with two day's work. (In the 1980s the period had grown to five months.) In the same year America exported to England 2½ million pounds of cheese (I hope it was not called swiss or cheddar then!) and 1 million pounds of tallow candles. None of Playfair's books or other writing records whether Jefferson replied.

Although Playfair can be credited with the invention of the modern methods of visualizing facts and figures he himself admits to a considerable influence from his brother. In the *Decline & Fall* of 1805 he acknowledges that "At a very early period of my life, my brother, who, in a most exemplary manner, maintained and educated the family his father left, made me keep a register of a thermometer, expressing the variations by lines on a divided scale. He taught me to know that, whatever can be expressed in numbers may be represented by lines."

One final note: A close inspection of his originals show that the "lines" are clearly "drawn" rather than "plotted" by joining up one point to another, and then another, and so on. Whether a preliminary sketch was made before the final engravings is not mentioned in any of the many notes, introductions, and descriptions that he wrote. Perhaps in the case of some early figures, as he admits "the materials (figures) . . . are not altogether accurate," he took the smoothed-out line that had to be used in those cases and continued it even though some of the rest of the chart may not have been well substantiated with statistics.

Playfair died in 1823.

Not the least remarkable thing about Playfair's work was that he was charting statistics at a time when there were few such facts and figures being collected in an organized way.

QUETELET CREATES THE NEW SCIENCE OF STATISTICS

The man who was recognized as the creator of the new science of statistics was Jacques Quételet (1796–1874). An office of statistics was set up in 1826 to take a census of what was then The Netherlands. In 1832 after the revolution, Quételet was asked to help, and the results of the population census of Belgium were published. He was made the

LEGISLATION

Of the following measures, which do you believe should be kept as they stand, which modified, which repealed?

KEPT MODIFIED REPEALED R DON'T KNOW

PER CENT
0 10 20 30 40 50 60 70 80 90 100

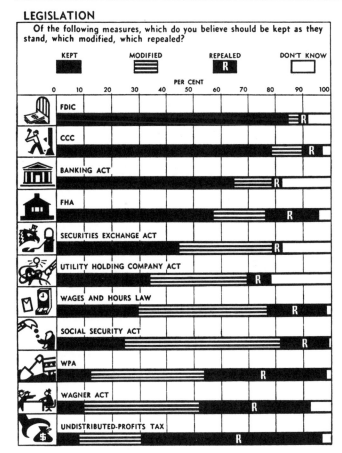

- FDIC
- CCC
- BANKING ACT
- FHA
- SECURITIES EXCHANGE ACT
- UTILITY HOLDING COMPANY ACT
- WAGES AND HOURS LAW
- SOCIAL SECURITY ACT
- WPA
- WAGNER ACT
- UNDISTRIBUTED-PROFITS TAX

LABOR UNIONS

Do you think business would be better off if the unions were to merge into one big powerful union? Would labor be better off?

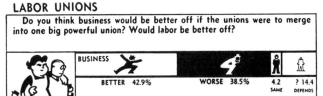

BUSINESS — BETTER 42.9% WORSE 38.5% 4.2 SAME ? 14.4 DEPENDS

LABOR — BETTER 53.7% WORSE 30.3% 4.7 SAME ? 11.3 DEPENDS

Do you think labor unions have helped or hurt this country as a whole?

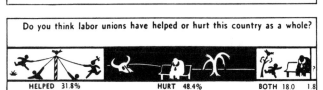

HELPED 31.8% HURT 48.4% BOTH 18.0 1.8

FOREIGN TRADE

In what one or two particular foreign countries do you see the best chance of building up an export market for your own products—or those of your industry?

LATIN-AMERICAN COUNTRIES 48.3%

What foreign countries give you the most competition in those markets?

GERMANY 40.6%

ENGLAND 30.3%

JAPAN 9.1%

INDUSTRIAL PROGRESS

What particular industry in recent years has, in your opinion, made the greatest technological progress?

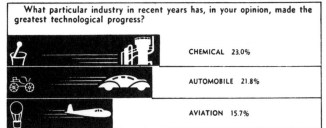

CHEMICAL 23.0%

AUTOMOBILE 21.8%

AVIATION 15.7%

What industry would you say has done the most to meet or go beyond the standards of performance that the general public expects of it, in order to win and deserve public approval?

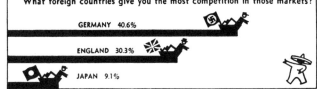

AUTOMOBILE 40.3%

RADIO 5.2%

RAILROADS 4.5%

What industry would you say had made the least technological progress?

TEXTILES 8.6%

RAILROADS 6.0%

BUILDING 5.1%

What industry would you say has, in its conduct, been apparently most indifferent to public opinion?

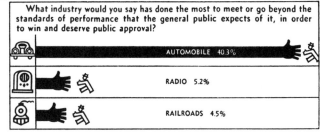

PUBLIC UTILITIES 9.9%

RAILROADS 8.2%

STEEL 7.0%

Charts by Irving Geis © Time Inc.

A whole page of charts from Fortune, *incorporating divided bars and ordinary bar charts, makes wonderful use of really inventive little symbols. The simplicity of the welcoming or shunning hands in the two charts at bottom right is graphically elegant and full of information. Modern chartmakers could learn a lot from the excellent use of blacks and whites in this example from almost 50 years ago.*

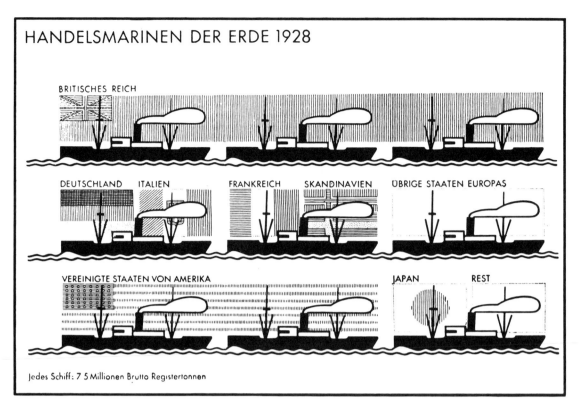

HANDELSMARINEN DER ERDE 1928

BRITISCHES REICH

DEUTSCHLAND ITALIEN FRANKREICH SKANDINAVIEN ÜBRIGE STAATEN EUROPAS

VEREINIGTE STAATEN VON AMERIKA JAPAN REST

Jedes Schiff: 7·5 Millionen Brutto Registertonnen

*Two charts from Otto Neurath's Vienna Museum of Social
and Economic Studies: (below) marriages in Germany
from 1911–1926 and (above) the world's merchant fleets
in 1928. It was Neurath's aim to give information visually
so that anyone could understand it regardless of language
or culture. He devised a picture language called "Iso-
type" (International System of Typographic Picture
Education) and applied its rules rigorously to his work.
The symbols (or isotypes) produced as part of the so-
called Vienna Method of representation remain models of
simplification and have exerted a very strong influence on
chartmakers from their invention in the 1920s to today.*

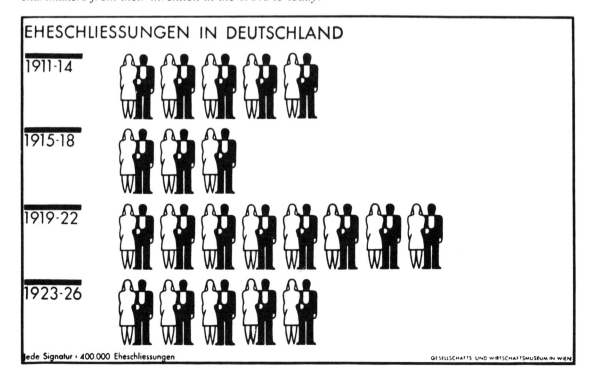

EHESCHLIESSUNGEN IN DEUTSCHLAND

1911-14

1915-18

1919-22

1923-26

Jede Signatur · 400.000 Eheschliessungen   GESELLSCHAFTS UND WIRTSCHAFTSMUSEUM IN WIEN

A pictorial bar chart comparing a year's worth of New York City waste (1922) to the height of the Woolworth Building (which was 29 feet higher). The obvious difficulty of determining how big an area on the ground should be covered by the base of the pile before it started rising (which could have resulted in a much higher pile had the ground area been smaller) is far outweighed by the visual impact of the idea.

President of the "Central Commission of Statistics" in 1841, and after much preparation he initiated the first international Congress of Statistics, which opened in Brussels on September 19, 1853.

His biographer, Naum Reichesberg, of the University of Berne wrote in 1896 of Quételet: "This vast spirit has rendered tremendous services as a mathematician. . . . His name has been written in golden letters, particularly in the fields of sociology and statistics." Quételet's 65 books and major papers on the subject led Emile Levasseur of the College de France to hail him in 1872 as a scholar who "first contributed to making statistics a moral science."

In America, charting did not really catch on until the late 20s. After the Wall Street crash, people became interested in the possibility of predicting the ups and downs of the stock market, at least in general terms, by using the historical data shown in charts. Whether or not there could ever be any real accuracy in this "history-repeats-itself" thesis, it marked the beginning of the regular use of charts in American publications.

In Austria, under the pioneering direction of Dr. Otto Neurath, the Vienna Museum of Social and Economic Studies produced in the 20s charts of precision, clarity, and graphic excellence that are still models of "how-to-do-it" today. In a publication describing the museum's work it was said: "If the immense power of statistical truth is to be turned to full account the prime necessity is the pictorial representation of statistical data. Our museum, with its carefully evolved method, is able, graphically, to represent social and economic problems. Meaningless columns of figures spring to life. Logic wedded to clarity is effective and convincing."

The "Vienna method," as it came to be known, had very strict principles. All irrelevant, decorative elements were banned. Symbols were drawn with the utmost simplicity. Instead of showing an increase in the quantity of a commodity by increasing the size of its symbol, the symbol itself was repeated as many times as was necessary to be representative of the required amount. By and large the drawing of objects in three dimensions was also avoided so as not to give a false impression of the quantity in question.

The museum charted the size of the world's armies, its production figures, commerce, population, religion, culture, and many of the symbols created by Neurath and the museum's art director, Gerd Arntz, can still be seen in charts today, for they have never been bettered.

This example of three-dimensional fever charts from West Germany in the 1950s gives information clearly and with nice symbolic touches. More tonal in execution than any of the other examples in the brief historical survey of this chapter, it was a forerunner of the more illustrated charts developed in the 1970s.

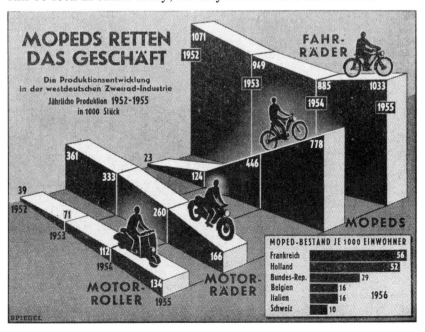

Chapter Two
The Four Types

STATISTICS MAY BE presented graphically in many different ways, but there should always be a sound reason for choosing the particular form of presentation. There are only four basic types of chart—fever chart, bar chart, pie chart, and table—and one of them will be the right form for any type of statistical information you may wish to display.

By and large it is the material itself that will determine which kind is to be used, for it will naturally be visually clearer in that form than in any of the others. The purpose of making a chart is to clarify or make visible the facts that otherwise would lie buried in a mass of written material, lists, balance sheets, or reports.

The type of chart chosen must enlighten the reader/user/viewer. It must seem to be the natural way for the material to be shown.

Taking each type of chart in turn, this chapter will ask six important questions about each one and will discuss and illustrate the differences:

1. What is the definition of a fever, bar, pie, or table?
It is important to know the differences among the types so that the best one is chosen for the job in hand.

2. What are its main elements?
It is important to fully understand the parts of each type so they can be put to the best use.

3. What variants are there?
It is important to be able to expand the scope of your work by being able to use one of the varieties of the four main types. To cope with any statistical problems you should know what varieties of the four main types are available.

4. What are appropriate uses?
It is important to know when it is right to use a fever chart, when a bar chart, when a pie chart, and when a table, and why each is successful.

5. What are inappropriate uses?
It is important to know when the wrong form has been chosen and why it is therefore unsuccessful.

6. What are the main requirements?
It is important to know what materials are needed to produce the chart, how difficult it will be, how long it will take, and what additional services and personnel may be required.

On the next four pages, a single typical example of each type will be shown; then the four types will be discussed in detail.

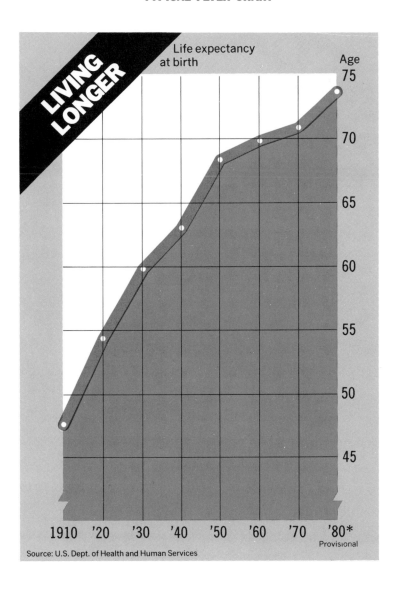

Source: U.S. Dept. of Health and Human Services

1. Definition
Visualization of quantities, plotted over a time period, by means of a rising and falling line.

2. Main Elements
Fever line, produced by joining together points plotted on a grid. (Otherwise known as "the curve.")

Y axis (vertical scale) represents quantities.

X axis (horizontal scale) represents time.

3. Variants
Single line.

Multiple line.

Three-dimensional line.

Combination with other types of charts.

Photographic treatments.

Illustrative treatments.

4. Appropriate Uses
Unemployment figures, financial time series, stock indices, or any set of figures the flow of which needs to be shown over a period of time.

5. Inappropriate Uses
Information too close together statistically to distinguish among the lines.

Too little variation in quantities along a single line.

6. Requirements and Considerations
Easy to produce for a fast look at the flow of a set of figures.

Graph paper for initial plotting.

Colored pencils needed to distinguish lines of different figures.

Minimal typesetting.

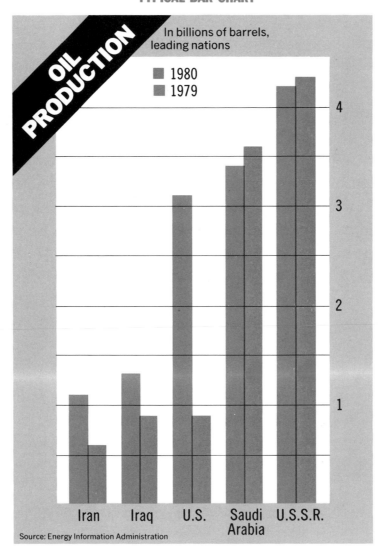

OIL PRODUCTION

In billions of barrels, leading nations

■ 1980
■ 1979

Iran   Iraq   U.S.   Saudi Arabia   U.S.S.R.

Source: Energy Information Administration

1. Definition
Visualization of quantities, each one represented by an individual bar or column corresponding in height or length to the amount being counted.

2. Main Elements
Bars, or elements grouped together in columns.

Grid, or a structural basis by which to understand the quantities.

3. Variants
Single row of abstract bars representing one commodity plotted over a time period.

Single row of abstract bars representing the values of different commodities all at the same time.

Three-dimensional versions of the above.

Bars as drawn representations of the commodity or subject described.

Small elements, each equivalent to one unit of the commodity being charted, graphically arranged into columns.

Multiple bars.

Double-ended bars: Both ends can be read against a scale.

4. Appropriate Uses
Prominence given to individual figures rather than to the overall flow.

Comparison of different commodities; not a time series.

To complement and show a difference between two or more sets of figures charted over the same time period.

5. Inappropriate Uses
Too many numbers, which would make the bars too thin.

Where the "flow" is more important than the individual numbers.

6. Requirements and Considerations
Take longer than fever charts.

Must be ruled out neatly to be convincing.

Typesetting plays an important role especially when different commodities or subjects are involved.

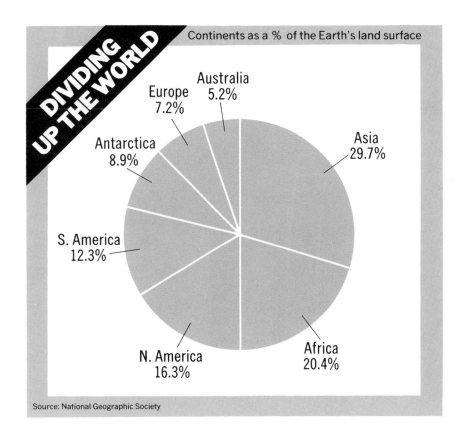

Continents as a % of the Earth's land surface

DIVIDING UP THE WORLD

Australia
5.2%

Europe
7.2%

Antarctica
8.9%

Asia
29.7%

S. America
12.3%

N. America
16.3%

Africa
20.4%

Source: National Geographic Society

1. Definition
The division of a whole into its components, usually in percentages.

2. Main Elements
Circle representing the complete number divided up by "spokes" from the center.

Segments formed as the radiating spokes touch the circumference of the circle, standing for the parts of the whole.

3. Variants
Flat, abstract two-dimensional.

Abstract three-dimensional.

As an element in an illustration or with illustrative elements attached to it.

As a total illustration, not necessarily circular, but still divided into proportions.

Divided bar, abstract.

4. Appropriate Uses
To show up to eight or ten component parts of a whole.

Budgets, share of market figures, analysis of income and spending.

5. Inappropriate Uses
Too many divisions, leading to impossibly small slices of the pie.

Too complicated an overall shape if it is not a circle.

6. Requirements and Considerations
Easily and quickly recognizable form.

Can be used very small with great effect by color coding the segments.

Careful, consistent typography important.

Compass, protractor, and percentage wheel essential.

QUALITY OF LIFE

U.S. metropolitan areas

	Per capita personal income[1]	Average price of a home[2]	Number of cloudy days a year	Crime rate per 100,000[3]	Percent jobless[4]
Atlanta	$9,294	$111,300	148	7,575	5.4%
Chicago	$10,455	$78,400	172	5,721	8.0%
Honolulu	$9,573	$104,900	101	7,574	5.8%
Indianapolis	$9,361	$62,700	175	6,009	7.8%
Los Angeles	$10,606	$118,800	91	8,418	6.6%
Miami	$9,714	$93,900	119	11,581	6.0%
New York	$9,839	$92,500	143	8,592	8.1%
Portland	$10,067	$97,400	227	7,324	7.8%
	[1]1979	[2]August 1981		[3]1980	[4]June 1981

Sources: Survey of Current Business, Federal Home Loan Bank Board, National Oceanic and Atmospheric Administration, Uniform Crime Reports, Bureau of Labor Statistics

1. Definition
Display of numbers or words arranged into columns.

2. Main Elements
The series of numbers and their subject titles.

A grid or framework to contain them.

3. Variants
Simple columns, with or without illustrations.

An illustrated frame that describes the subject matter.

A photographic background.

4. Appropriate Uses
Family trees, timetables, flow charts, distance/mileage charts, and calendars.

Comparisons of numbers that are too great in spread to be easily charted.

Where the exact numbers must be read, rather than illustrated as a generalized flow.

5. Inappropriate Uses
Where it is possible to plot the statistics as a chart. Don't miss the chance!

5. Requirements and Considerations
Type is very important: It is the main element and must be readable.

THE FEVER CHART

Together with the bar chart, the fever is the most often used form for the pictorial representation of statistics. The fever chart is particularly suitable for pinpointing accurately a given quantity on a given date, but it also shows the generalized trend in a set of statistics: that is, the "flow" of the figures in historical terms and an indication of the future, if projections or estimates are included. Much can be learned from the close study of a fever chart; indeed there are stock market analysts who specialize in future projections of stock performance based on such study.

The fever chart can best be understood by examining it in light of the six questions introduced at the beginning of this chapter.

1. What Is the Definition of a Fever Chart?

The fever chart is named after the piece of paper hanging on the end of a hospital patient's bed that traditionally plots his or her rising and falling temperature. Sometimes it is referred to as a "line graph."

It is a visualization of quantities, plotted over a period of time, with both the quantities and the time shown together. An example is the rising and falling of inflation over a period of years, or the change in the number of passengers on airlines.

2. What Are Its Main Elements?

First and foremost is the line from which this form gets its name. This is produced by joining together points plotted on two axes. In general the time period is plotted along the x, or horizontal, axis (abscissa), and the quantity being charted is plotted up (and down) the y, or vertical, axis (ordinate). The x and y axes are at right angles to one another. For complete accuracy in the reading of the fever chart, the line should be drawn by making a series of short straight lines from plot point to plot point. However, if the plot points are clearly marked (by small dots, for instance) the line joining them may be smoothed out, thus producing a more generalized "curve" that shows at a glance the direction and trend of the statistics being charted.

The horizontal scale is generally labeled at the bottom of the chart. So that there is no distortion of the information, the time divisions must be equally spaced along this axis, even if, at the time of plotting, information is not available for, say, each consecutive year. If some years are missing, there must nevertheless be a grid line for that year, and the fever line should be dotted or even left blank for that year only and should start again at the next year for which data are available.

The vertical scale, indicated along the y axis, usually represents a series of quantities or percentages of the subject being charted. As is the case with the horizontal scale, it is one of the grid lines and can be shown at either/or both sides of the chart as a slightly heavier line than the other grid lines.

The grid is formed by extending lines vertically or horizontally from the points at which the measurements from the horizontal and vertical axes are marked on their respective scales. If the graph is drawn on graph paper, all the lines are preprinted, and it is a question of deciding which of the lines already there are to be used. For the final production of the chart most of these lines will be omitted, so that the fever line itself stands out clearly. Just enough grid lines must be drawn to enable the reader to understand the information. But where only a quick general impression of the statistics is intended, the grid lines can be kept down to a minimum, and in some cases, only the horizontal and vertical scale lines, with their attendant numbers, are all that is needed.

The plot points mark the intersection of time and quantity on the grid and form the skeleton for the line itself.

Labels announce to the reader what the chart is about, denote the amounts and time period involved along the vertical and horizontal scales, and clarify the unit of measurement (for instance, where "unemployment" may be the subject of the chart, "percentage of civilian labor force unemployed at year-end" clarifies the unit of measurement). It is very important to provide essential details, such as whether part of the plotting is estimated or projected; to note the difference between lines, where more than one item is shown; and to give a source from which the original information was obtained. Although some of these labels will, and should, be small, they are nonetheless essential parts of a good chart, for without them the reader only gets part of the story.

If it is neater to do so, use an asterisk with a note at the bottom of the chart to clarify a point in the title or elsewhere. Sometimes a fuller explanation is needed than can be neatly fitted where it occurs on the chart. Labels ensure that the reader is left in no doubt about what any part of the chart means, and misleading or incomplete labels are just as annoying to a reader as misleading, incorrectly placed, or missing signposts are to the traveler wishing to find his or her way around an unknown area.

Color, when available, helps to differentiate lines in a multiple fever chart. Obviously, it can also liven up any, even the most simple, fever chart, but it should never be used for decorative purposes only.

Where it is not possible because of lack of space to label each line separately in a multiple fever chart, a key should be used, showing a small sample of each type or color of line, with the appropriate description next to it.

3. What Variants Are There?
The fever chart may take the form of one rising and falling graph line (the "curve"), or it may be a number of these lines standing for different subdivisions of the subject being graphed.

It may be projected into three dimensions and stand by itself as a solid shape, or it may be made into part of a picture (like a mountain range or a roller coaster). It might also be combined with another form of chart, for instance, the fever line tracking one batch of numbers and a bar chart another, both over the same period of time.

This graph of declining truck traffic is an example of a single line chart, with the grid showing above and/or below it. It is also called a "simple" line graph because a single set of values is connected by one line. The side of the truck is drawn parallel to the picture plane as a frame for the chart.

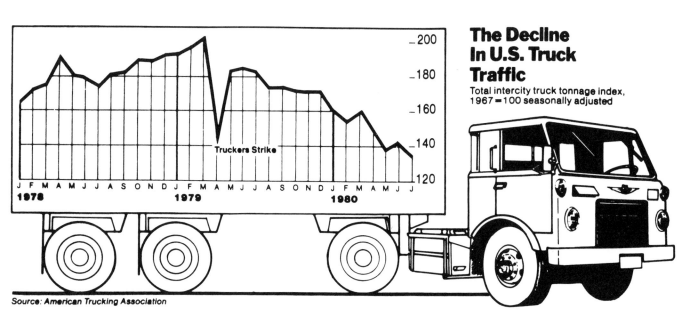

The Decline
In U.S. Truck
Traffic
Total intercity truck tonnage index,
1967 = 100 seasonally adjusted

Truckers Strike

200
180
160
140
120

J F M A M J J A S O N D J F M A M J J A S O N D J F M A M J J
1978   1979   1980

Source: American Trucking Association

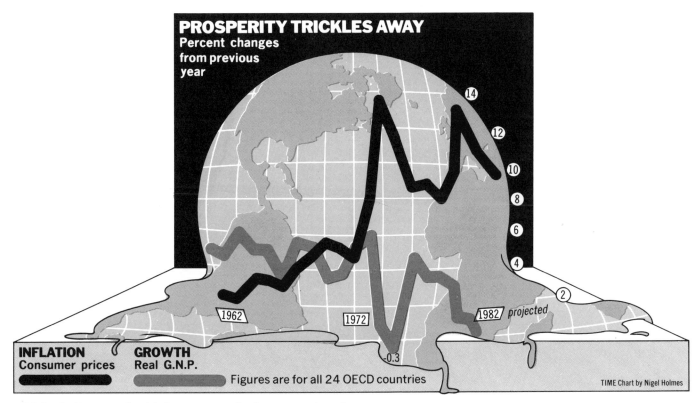

PROSPERITY TRICKLES AWAY
Percent changes from previous year

14
12
10
8
6
4
2

1962   1972   1982 *projected*

-0.3

INFLATION
Consumer prices

GROWTH
Real G.N.P.

Figures are for all 24 OECD countries

TIME Chart by Nigel Holmes

Multiple lines, each with its own set of values, are plotted on the same scale. Here inflation and growth are plotted together since they both fall within the same percentage range. Although they are different things, what is measured here is the percentage change from year to year in both of them, not actual quantities.

The grid lines do double duty here as conventional graph paper on which to plot the points and also as a reminder to the reader of longitude and latitude lines on the globe in the background. In this sense the globe is different from the truck in the chart on the opposite page, as it is both the background for the chart and an illustration of what the chart is about. The link is the grid lines, which naturally appear on graph paper and on a globe.

Note the stepped fever line. Certain interest rates (for example, the prime rate) are not adjusted monthly, but remain in force until the day a decision is made (by a majority of banks, in the case of the prime rate) to change them. On that date the rate changes up or down to a new rate and remains there until a new rate is decided upon. There is not necessarily any regularity to these changes, which accounts for the particular look of this variant of the fever chart.

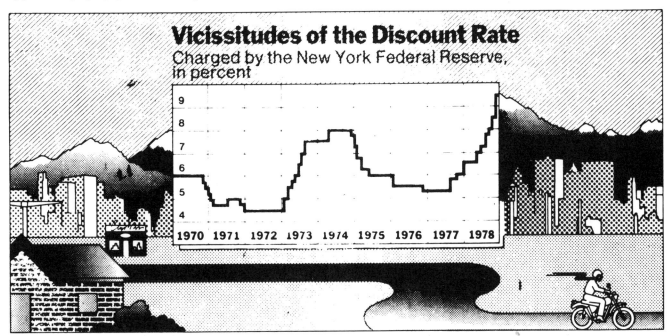

Vicissitudes of the Discount Rate
Charged by the New York Federal Reserve, in percent

9
8
7
6
5
4

1970   1971   1972   1973   1974   1975   1976   1977   1978

Here the interaction of the humans with the otherwise abstract, three-dimensional fever chart illustrates the point that the numbers themselves make. The amount of cash available for Social Security payments is a solid fever line when it is above the zero line but becomes a hole when the cash runs out and the numbers go into deficits, swallowing up the poor people who may no longer receive their payments. The chart is drawn in perspective and then tilted to give a disoriented and destabilized feeling to the whole image.

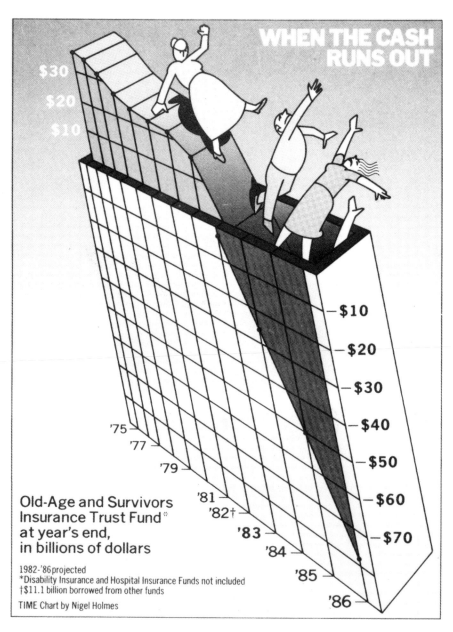

WHEN THE CASH RUNS OUT

$30
$20
$10

—$10
—$20
—$30
—$40
—$50
—$60
—$70

'75
'77
'79
'81
'82†
'83
'84
'85
'86

Old-Age and Survivors Insurance Trust Fund* at year's end, in billions of dollars

1982-'86 projected
*Disability Insurance and Hospital Insurance Funds not included
†$11.1 billion borrowed from other funds
TIME Chart by Nigel Holmes

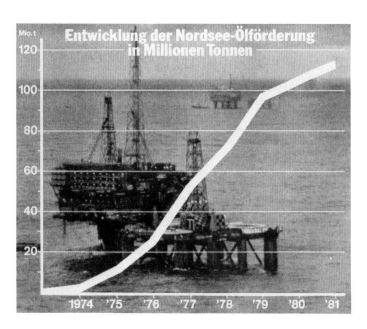

Many times it's appropriate to use a photographic background. Here a simple fever line with grid lines and labels is dropped out in white from a photograph of a North Sea oil drilling platform. This gives a very fast visual impression of the subject.

Three-dimensional line forms the top of the load on the tractor trailer and swoops down dangerously towards the poor farmer. Here the plot points are marked, but the overall trend down is outlined in a wider stroke.

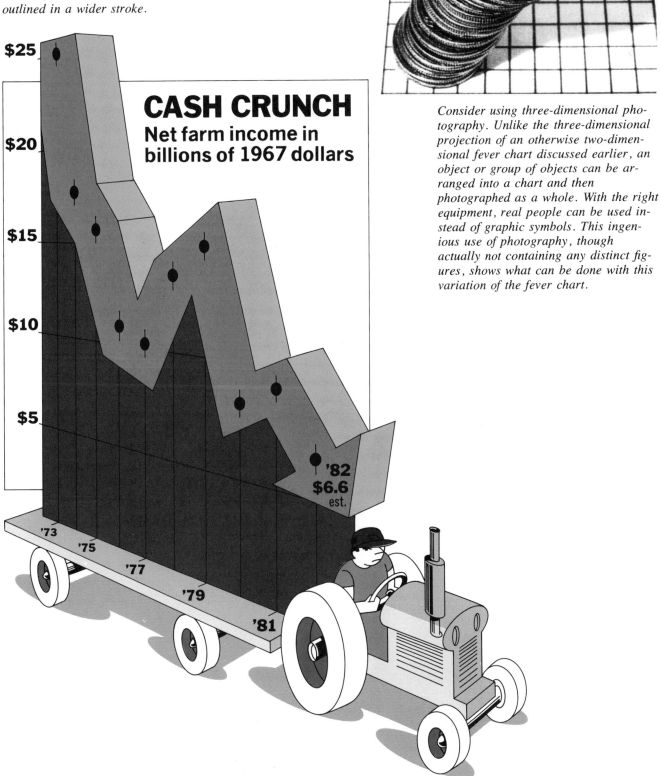

CASH CRUNCH
Net farm income in billions of 1967 dollars

$25

$20

$15

$10

$5

'82
$6.6
est.

'73

'75

'77

'79

'81

Consider using three-dimensional photography. Unlike the three-dimensional projection of an otherwise two-dimensional fever chart discussed earlier, an object or group of objects can be arranged into a chart and then photographed as a whole. With the right equipment, real people can be used instead of graphic symbols. This ingenious use of photography, though actually not containing any distinct figures, shows what can be done with this variation of the fever chart.

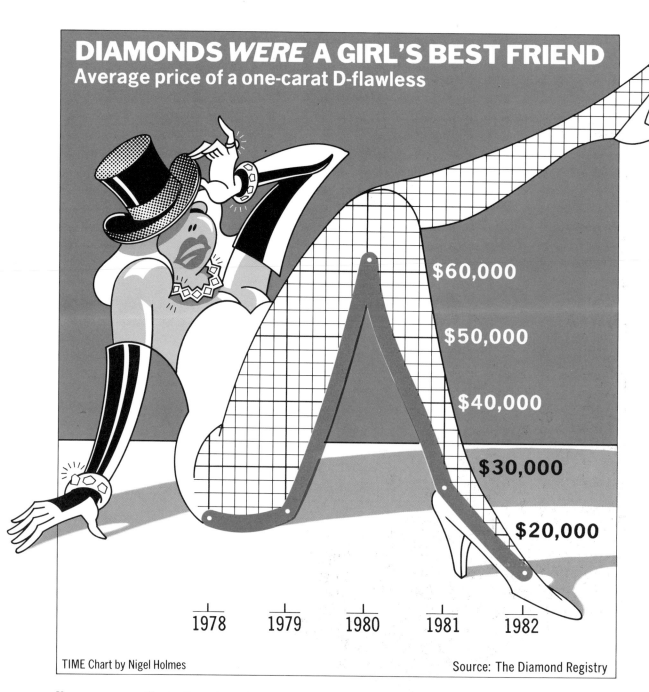

DIAMONDS *WERE* A GIRL'S BEST FRIEND
Average price of a one-carat D-flawless

$60,000

$50,000

$40,000

$30,000

$20,000

1978   1979   1980   1981   1982

TIME Chart by Nigel Holmes

Source: The Diamond Registry

You can use an illustration where the actual shape of the plotted line can, without being forced, become a part of a drawing itself. More obvious cases might be mountain ranges and roller coasters, which in themselves can move up and down easily and thereby visually describe an idea while at the same time present the figures.

4. What Are Appropriate Uses?

The successful fever chart is one that shows clearly the relationship of the point immediately before it and the one immediately following it so that the reader can understand the flow of the figures. The line that results from connecting the various points can then be shown by itself or in combination to compare two or more sets of statistics.

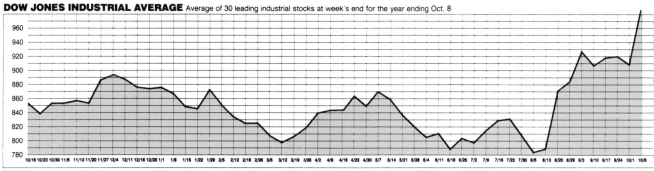

DOW JONES INDUSTRIAL AVERAGE Average of 30 leading industrial stocks at week's end for the year ending Oct. 8

TREND: Over the past year, the Dow Jones index of 30 industrial issues at week's end has ranged from a high of 986.85 on Friday, to a low of 784.34 on Aug. 6, 1982.

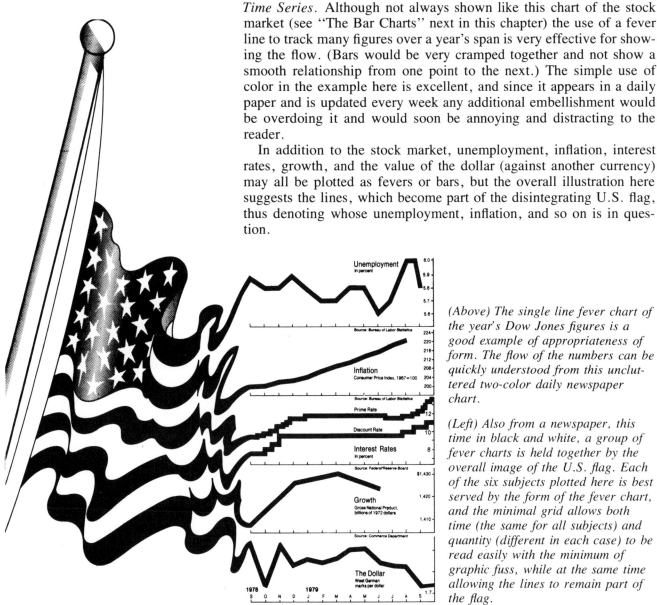

Time Series. Although not always shown like this chart of the stock market (see ''The Bar Charts'' next in this chapter) the use of a fever line to track many figures over a year's span is very effective for showing the flow. (Bars would be very cramped together and not show a smooth relationship from one point to the next.) The simple use of color in the example here is excellent, and since it appears in a daily paper and is updated every week any additional embellishment would be overdoing it and would soon be annoying and distracting to the reader.

In addition to the stock market, unemployment, inflation, interest rates, growth, and the value of the dollar (against another currency) may all be plotted as fevers or bars, but the overall illustration here suggests the lines, which become part of the disintegrating U.S. flag, thus denoting whose unemployment, inflation, and so on is in question.

(Above) The single line fever chart of the year's Dow Jones figures is a good example of appropriateness of form. The flow of the numbers can be quickly understood from this uncluttered two-color daily newspaper chart.

(Left) Also from a newspaper, this time in black and white, a group of fever charts is held together by the overall image of the U.S. flag. Each of the six subjects plotted here is best served by the form of the fever chart, and the minimal grid allows both time (the same for all subjects) and quantity (different in each case) to be read easily with the minimum of graphic fuss, while at the same time allowing the lines to remain part of the flag.

33

A combination fever chart and bar chart is used here to link together two different subjects with two different scales. The factor common to both is the time period, but the fact that one scale is percentages and the other thousands of units demands that different chart forms be used to clearly differentiate the information and play one sort against the other, while retaining the link between them.

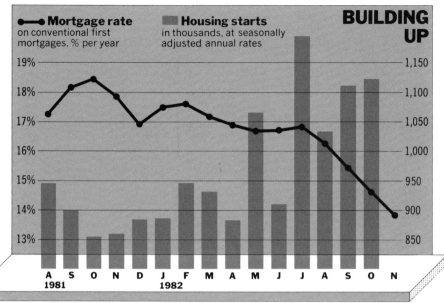

Combination of Information. Where one sort of information is to be played against another in a related field, but counted in a different manner, a fever line can be used effectively. Here a mortgage rate is shown in relation to the rate of building new houses ("housing starts"). The correlation between high mortgage rates and low building activity and vice versa can be clearly seen.

5. What Are Inappropriate Uses?

If there is too much information to be dealt with, or if the scale jumps from a tiny amount to a huge one, or if there is very little variation in the quantities, then the fever chart is not the form to use. Sometimes a great deal of plotting will be completed before it becomes clear that the lines are all going to cross over one another, thus possibly obscuring the information. At the beginning, the chartmaker will have to go through this exercise, but with experience and a careful preliminary reading of the data, it will be possible to save time by recognizing that a different form, for instance, a bar chart or a table, would better serve the information.

The simple rule is ease of reading. If you cannot be objective enough about whether the information is decipherable in your chart, ask a colleague whether or not it is.

Since the appropriateness of the fever chart has already been established for depicting a time sequence set of statistics, it follows that this form should not be used where no time period is involved. By joining one plotted point to its neighbor, there is an automatic implication that the second point has actually followed the first, as one event follows another. You will see, in the case of bar charts, that a number of different commodities can be plotted for the same time period. It would be nonsense to join all the plot points up as though they were connected.

If there is too small a variation in the quantities—say, the same number with different decimal point variations—then a table of the figures is probably a better choice. Some kinds of information need to be stated to two or three decimal places. Where this is the case and each figure has to be read, it is certainly better to print the figures themselves as a table, rather than plotting them on what would necessarily have to be an extremely complicated and finely ruled grid.

For examples of inappropriate charts, see Chapter 6.

6. What Are the Main Requirements?

In its most basic form, the fever is an easy chart to produce. An artist can draw a passable chart in twenty minutes simply by drawing a grid, or using tracing paper over already printed graph paper, plotting the points from the two coordinates, and joining them up to form the fever line. However, it is important to know who is going to read your chart, for this will determine the final degree of finish: whether or not you write in the labels and numbers by hand, by stencil lettering, by using a burnish-on type, or by having the type set by machine.

Materials. Graph paper, which is available in different colors and with different increments per inch, should be kept in stock at all times. It will save you hours of boring work ruling grids. Tracing paper is the natural companion to graph paper. It allows the artist the chance to see the whole grid while plotting, but only end up with the most important lines actually traced through. This simplifies the chart by eliminating unnecessary grid lines—and saves graph paper too!

Some tracing pads are supplied with a sheet of graph paper bound in. This is often printed on heavier weight paper than normal and may be divided into eighths of an inch on one side and tenths on the other. This is very useful, but does not eliminate the necessity of having your own separate supply of paper. Where plotting is very detailed, it should be done directly onto a new piece of graph paper, so as to stop the chance that your tracing paper might slip, be moved inadvertently, or, by its very nature, slightly obscure the grid underneath.

Colored and lead pencils always kept sharp will be needed for plotting. Use 2H or harder so that really fine, clear lines are drawn—even though it may be your intention later to draw a thicker fever line. By plotting properly in the first place, you will have an accurate base on which you may later choose to draw a more generalized curve.

You will also need a ruler with measurements in picas and inches, as well as 45-degree and 60-degree triangles, T-square, eraser, compass, Scotch tape, and register marks. These basic materials will see you through the initial stages of all chartmaking. For finished art you will also need a set of technical pens (with a compass attachment) in as large a range of sizes as you can afford. Buy the "jewel point" (or similar) for the thinnest sizes. The lines drawn by cheaper ones will thicken with use, so that it will become impossible to match line widths if you go back to the job later for corrections or additions. Make a note on your final tracing of the pen you have used for drawing each of the different kinds of line employed in the final artwork.

Trial and error will teach you which ink to buy for your pens. A pen that does not work because of clogged ink is infuriating, but whether or not they do clog depends on how often you use them, the weather—or atmospheric conditions—and on their own good or ill nature! Few of the "cleaners" on the market are as good as old-fashioned water and a little patience. Remember that after flushing a pen through with water, the line drawn will be paler to begin with, as it is virtually impossible to shake out all the water and thus the black ink is diluted for a short while. Scribble with the pen until it produces a jet-black line.

The material most generally used for finished art is Mylar or frosted prepared acetate. This can be bought in different weights and sizes, and again it is a matter of personal choice. Some of the makes are more transparent than others, but they can all be fairly safely relied upon not to stretch or shrink, unless subjected to extremes of temperature. Do not be tempted to dry a large area of black ink under an adjustable lamp pulled down to within an inch of it. It will dry it, but it will also curl the material so that it does not lie flat and is therefore hard to register with other overlays.

Technique. The descriptions that follow, together with those above on materials, apply to all forms of chartmaking and will only appear here.

Tones, available in a large range of dot sizes (screens), can be applied directly to line artwork by cutting areas from sheets of adhesive-backed film. The 55-line screen—that is, 55 dots to the inch—is a fairly coarse screen, for use in a newspaper. The screen size used for tones in this book is a 133-line screen.

Drawing. Drawing, in this context, refers to the way the idea behind the numbers becomes visible on the page. The fever line is plotted and drawn roughly by hand or very accurately with a ruler in pencil so that mistakes can be rectified and changes incorporated. It is important that everything about the chart be worked out at this early and easily modifiable stage, so that at the next stage, drafting, the work can proceed without hesitation or change.

Drafting. Drafting should be the merely mechanical procedure of turning the drawing into a printable form—a method of making pencil lines, with all their attendant erasures, notations, and wrong beginnings, into a neat, even, solid line that is thoroughly intentional. Good drafting is essential to the process of completing a good chart, for it is that part of the job which will eventually end up in the hands of the printer and which therefore most nearly represents the final printed image.

Typesetting. There are two cases where there may be no typesetting: first, if the chart is simply to be shown to a limited audience and not to be published; second, if there is so little type that it could easily be added with a pressure-sensitive, rub-down product. Most manufacturers have approximately the same range of typefaces, and they include those that are the most obvious choices for charts: Helvetica, News Gothic, Franklin Gothic—the straightforward sans serif faces. However, it is very difficult to keep the type really straight using these products for more than a few words, especially at the small sizes often used in charts, and so the best results can be obtained by using typesetting of one sort or another.

Some typewriters are equipped with perfectly acceptable faces, though limited in size, but the best results will always be obtained by having the type set on a professional typesetting machine. Since this is possibly the only part of the production of your artwork that you do not actually do yourself, plan ahead for the type so that it does not hold you up at the end of your job. It should be possible to work out the size of type needed at the same time as you plan the rest of the chart.

Whenever possible, be involved with the application of the type to the chart. In a freelance situation, art directors may suggest that they handle the type—you just do the charting. They are missing the point of the chart, which by its nature contains numbers and type that are part of the design. A chart is not, and should not be thought of as, an illustration, which can be manipulated later in an art department. If the numbers are not ready for you to apply to the chart, then you should not be starting the chart at all.

Tone. Tones of black and white are used to differentiate areas or types of information. Thus a background may be a light tone and the fever line dark to contrast with it, or vice versa. If more than one line is plotted, a different tone may be used for the second and subsequent lines or areas between them. Tone can be applied by cutting Amberlith overlays, which, duly noted, are translated into screens by the printer. Alternately, it may be applied directly to the artwork by cutting areas of it from sheets of pressure-sensitive film, which are manufactured in various screen sizes and applied as a line constituent of your artwork (since the printer does nothing to it but treats the entire artwork as a piece of line).

Remember to use the correct screen size for the type of reproduction and for the type of paper on which the chart will be printed [rule of thumb: no more than 55 lines (dots) to the inch for newspaper or newsprint reproduction]. If you are working larger than the final printed size, allowances must be made for the reduction involved. By applying

(Above) The different patterns available in adhesive-backed film can literally illustrate such things as bricks, grass, and leaves, or be used to great effect as abstract tonal areas in a drawing.

your own tone as line, you are also able to control the look of the image in an interesting way. If you deliberately choose to use a screen of larger dot size than would usually be used in a straightforward toned conversion, a whole set of different visual effects are possible. You can also experiment with different patterns as areas of tone instead of simple dots.

Color. When the budget allows, color can be used instead of, or in conjunction with, tones of black and white, as described above. Sometimes only one color is available. If you have a say in the matter, red is generally the best choice, unless the subject itself suggests another color—for example, blue for a chart about water shortage or yellow for gold prices. Except in such obvious cases, red as a second color is the most versatile in terms of tonal value and simple graphic clarity.

The accompanying illustration shows the wide variety of colors that can be made from mixing just red and black of different tonal strengths.

Where tone is not involved, red consistently is the brightest and clearest color for adding emphasis to a chart. Parts of this book are produced in black and white, parts in full color, and parts in two color. There was never any doubt in my mind about the choice of color for the two-color sections. Red simply does the best job on all counts: versatility of tonal variation, graphic clarity, most straightforward signaling of a point to be emphasized, and greatest contrast with black.

Where full- or four-color is available for an artist to use, care should be taken that it is not used too decoratively, simply to dress up the information. Impact may be achieved by limiting yourself fairly strictly. In print the job will probably be in full-color only because color pictures are being used elsewhere on the page or magazine, not necessarily because it is deemed essential that the chart be in full color. Do take the opportunity when it arises, but use the color carefully.

As with tone, color can be indicated on your artwork with overlays. Although this is technically the best method—certainly the method that will produce the finest reproduction of your work—be prepared to supply a finished colored rendering of your chart along with the final art. Few clients will be able to understand, let alone visualize, from the appearance of your final art with its overlays instead of the actual colors the ultimate look of the chart when it is printed. Do not give a client any chance to reject your hard work because they do not understand the technicalities of print preparation. There is no reason why they should understand it. The added work on your part of preparing a tissue overlay to indicate what the colors will look like will pay off in the end.

(Right) A printed combination of just two colors—red and black—produces a huge range of tonal values and colors. Each block in the left-hand column is 10% black, overprinted with increasing tonal intensities of red. Thus, from the top, the red blocks are 10%, 30%, 50%, 70%, and finally 100% in the bottom left-hand corner. The second column from the left is 30% black, the third 50% black, the fourth 70% black, and the final column, on the right, is 100% black. The same sequence of red tones as was used in the right-hand column is overprinted in the other columns.

THE BAR CHART

This form of chart lends itself to many variations and is in some cases interchangeable with the fever chart, but while that is most concerned with the representation of a time series, the bar is also suitable for the depiction of non–time-related subjects.

Let's examine the bar chart in relation to the following six questions.

1. What Is the Definition of the Bar Chart?

The bar chart is a series of bars or columns representing amounts of data. The height or length of the bar equals the amount being shown. Some definitions of a bar chart call upright bars "columns" and horizontal bars "bars."

2. What Are Its Main Elements?

Elements grouped together into bars or piles or columns are the primary ingredients in a bar chart. They may be laid out vertically or horizontally or at some angle in between. In order to understand the quantity of the commodity or subject being measured, it must be laid out according to a grid or scale, which is then an essential part of the finished chart. Color and labeling are other main elements, as in the case of fever charts.

3. What Variants Are There?

Like fever charts, bar charts come in many different forms—from simple upright lines, through the third dimension and illustration, to in-scale comparisons of actual objects.

Source: Annual USTA Yearbook for 18 under boys' rankings.

Photographic bar charts use real objects or people instead of drawn ones to depict the statistics.

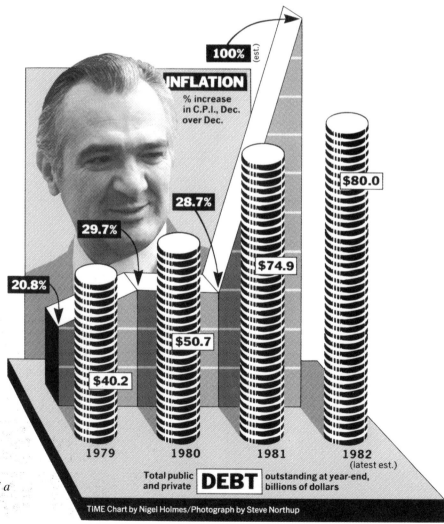

Photographs may be combined with other graphic forms, as in this case of a three-dimensional bar chart and a three-dimensional fever line.

Total assets up 75% since 1976

*1978 results adversely affected by a strike at The New York Times

1980	$449,546,000
1979	$389,786,000
1978	$316,806,000*
1977	$297,018,000
1976	$256,543,000

The simple bar chart is a flat, two-dimensional abstract drawing. In this illustration from an annual report, the most basic type of bar is used, and since the exact quantities need to be read and understood, they are all printed at the end of each bar. Note that the expected onward flow of assets was stopped by a strike in 1978, a fact that the author felt was necessary to explain. There is no particular reason to use a bar chart in this case, since it is a simple time-related progression of numbers, nor was it essential that the bars be horizontal, as opposed to vertical.

The bars themselves may be drawings of the commodity being measured in the chart— in this case money.

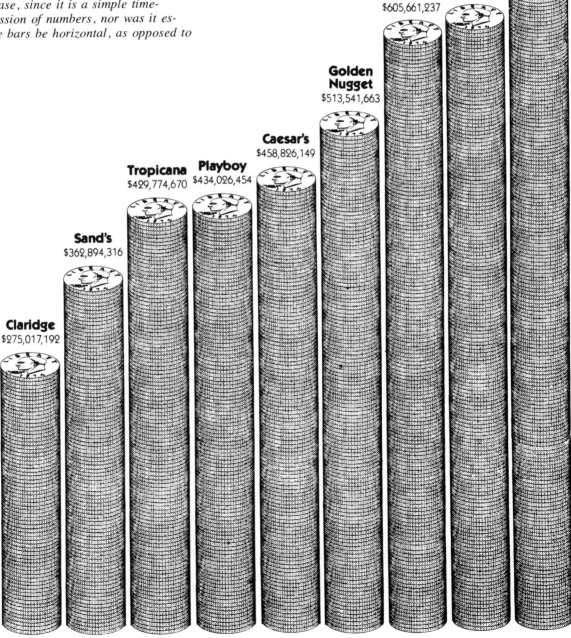

Claridge
$275,017,192

Sand's
$362,894,316

Tropicana $429,774,670

Playboy $434,026,454

Caesar's
$458,826,149

Golden Nugget
$513,541,663

Bally
$605,661,237

Resorts
$620,511,787

Harrah's
$674,362,570

Double-ended bars are useful for work-in-progress flow charts and, as in this case, for plotting the three points in a stock market day: the high, the low, and the close.

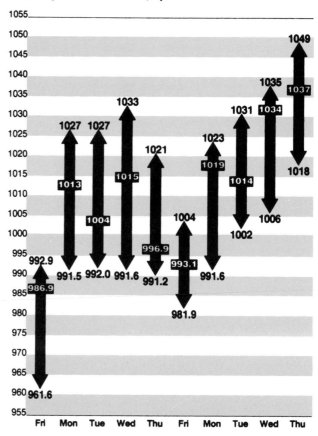

DOW JONES INDUSTRIAL AVERAGE
Average of 30 leading industrial stocks
Yesterday's close: 1036.98, up 2.86

A simple, flat drawing may be projected into the third dimension while still remaining abstract. The artist has neatly used the black area at the top of each bar to contain the exact quantity of each bar, thus providing precision with a simple visual depiction of the quantity. The title box and secondary (fever) chart are treated with a similar black shadow to give a cohesive look to the whole.

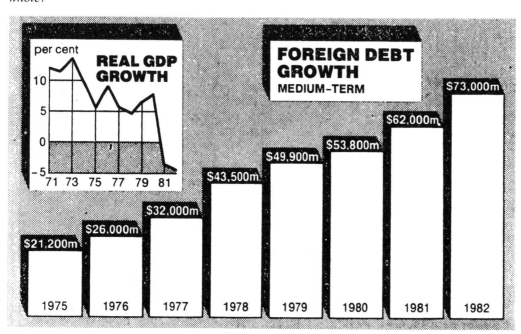

Catches by Major Fishing Country (1979)

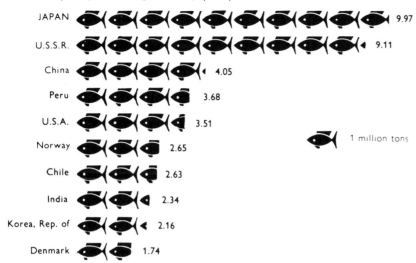

JAPAN	9.97
U.S.S.R.	9.11
China	4.05
Peru	3.68
U.S.A.	3.51
Norway	2.65
Chile	2.63
India	2.34
Korea, Rep. of	2.16
Denmark	1.74

🐟 1 million tons

Bars may be composed of a number of individual elements, each one representing a certain amount of the substance being charted. This is a very useful and highly recommended form of chart. By keeping all the elements the same size, the common and misleading error of changing the size of one element—in this case one fish—to represent the different quantities is avoided. Somewhat of a drawback, however, is the problem of how to deal with fractions of the whole. Here the exact amount is printed so there is no confusion.

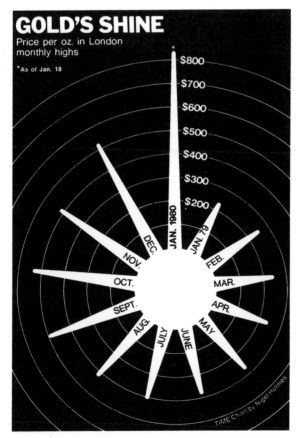

Circular bar charts, sometimes called "star graphs," are particularly useful for showing the first figure to be plotted immediately next to the last one. In this chart of gold prices, the beginning of one year is plotted next to the beginning of the following one.

Famous Mountains

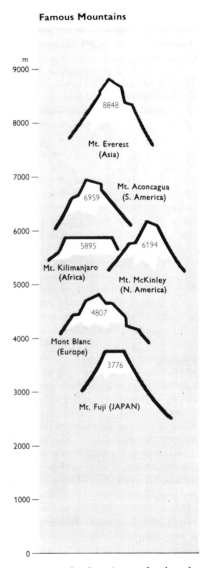

In-scale drawings of related objects can be called "pictorial" bar charts of a kind. They are visual comparisons after all. These simplified drawings of the highest mountains in the world could just have easily been drawn as simple bars—but how much more effective they are as the shapes of the mountains themselves.

41

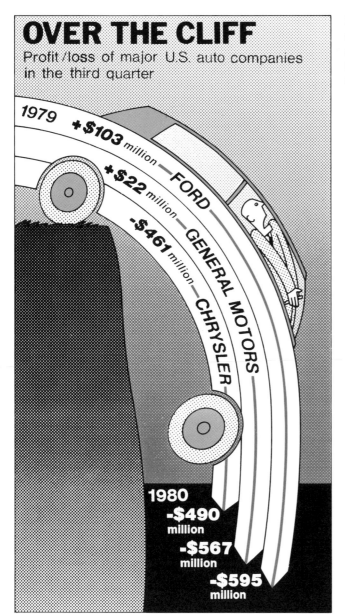

OVER THE CLIFF
Profit/loss of major U.S. auto companies in the third quarter

1979
+$103 million — FORD
+$22 million — GENERAL MOTORS
-$461 million — CHRYSLER

1980
-$490 million
-$567 million
-$595 million

The bars may be bent into a shape that in some way describes or amplifies the data. Beware of the possibility of distortion. This example shows the exact amount in figures.

The bars may be projected down instead of up, especially when doing so is appropriate to the subject. In this instance, the idea of being deeper in debt or trouble is matched by the downward thrust of the bar/shadow.

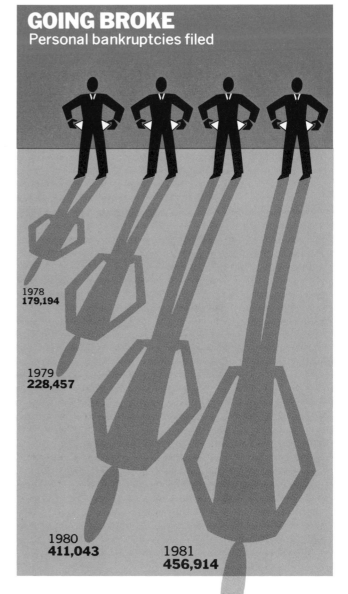

GOING BROKE
Personal bankruptcies filed

1978
179,194

1979
228,457

1980
411,043

1981
456,914

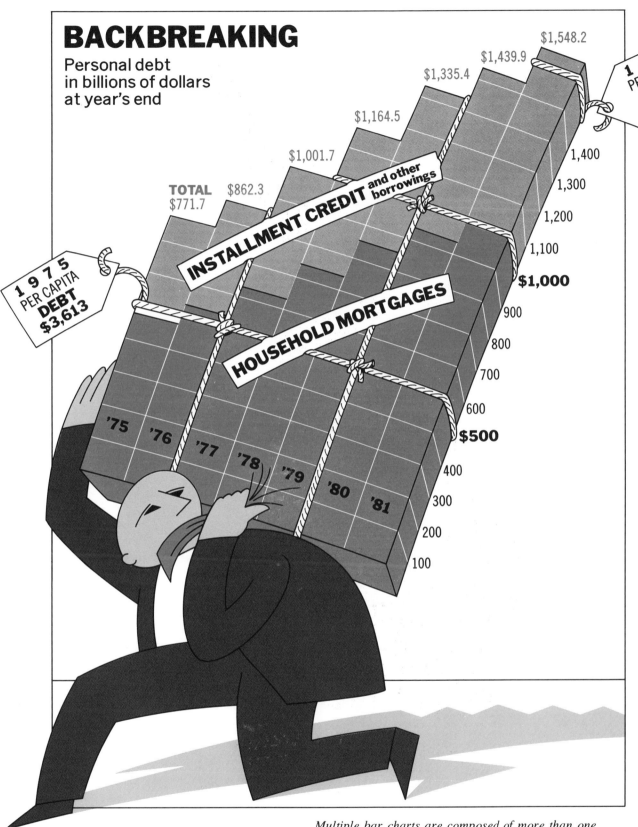

BACKBREAKING

Personal debt
in billions of dollars
at year's end

$1,548.2

$1,439.9

$1,335.4

$1,164.5

$1,001.7

TOTAL $862.3
$771.7

1 9 8 1
PER CAPITA
**DEBT
$6,737**

1 9 7 5
PER CAPITA
**DEBT
$3,613**

INSTALLMENT CREDIT and other borrowings

HOUSEHOLD MORTGAGES

'75 '76 '77 '78 '79 '80 '81

1,400
1,300
1,200
1,100
$1,000
900
800
700
600
$500
400
300
200
100

*Multiple bar charts are composed of more than one
ingredient. Two elements for the same time period are
placed one on top of another. At the most only three
ingredients should be used for each bar, as it becomes
very difficult to read the quantities of any but the bottom
one—that being the one which starts at zero—without a
fair amount of mental arithmetic.*

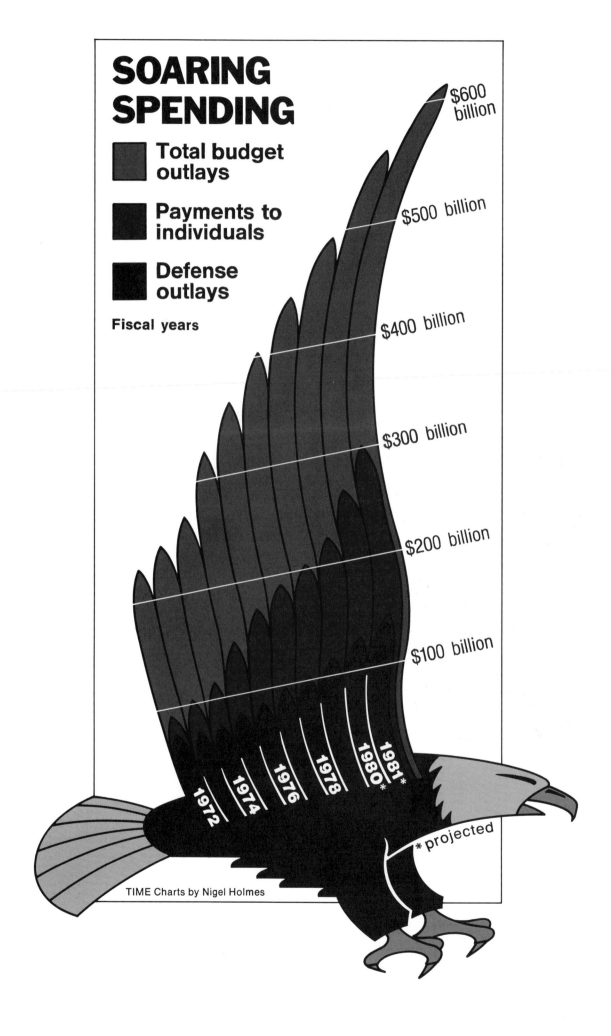

TIME Charts by Nigel Holmes

MONSTROUS COSTS

Total House and Senate
campaign expenditures,
in millions

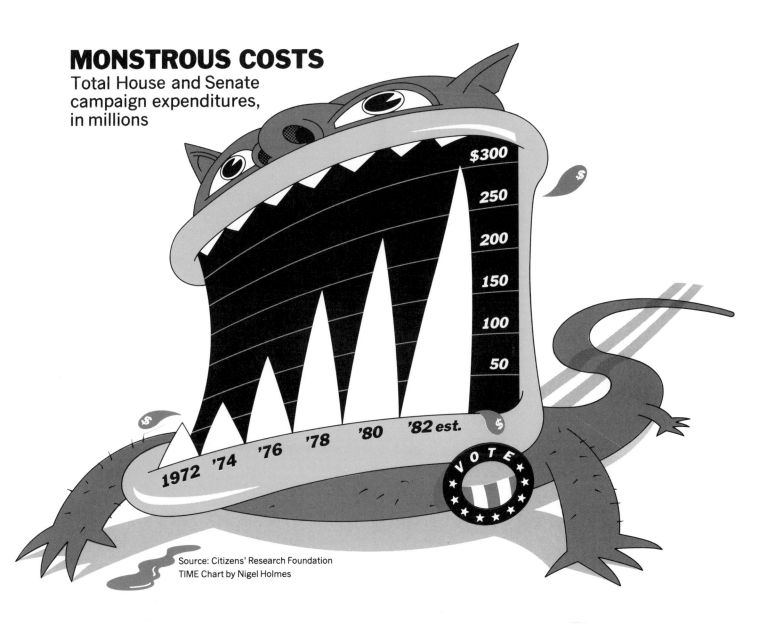

$300
250
200
150
100
50

1972 '74 '76 '78 '80 '82 est.

VOTE

Source: Citizens' Research Foundation
TIME Chart by Nigel Holmes

*In certain cases the measured elements
in a bar chart can be woven into a total
picture so that the whole drawing is a
chart. These two animal charts demonstrate this principle: No abstract
elements are added to the eagle or the
monster. They are not backgrounds or
adjuncts to a chart: They are the charts.*

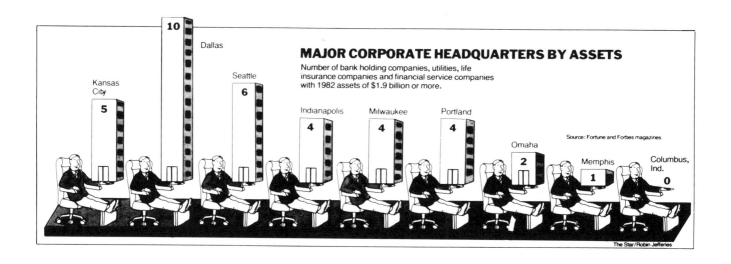

MAJOR CORPORATE HEADQUARTERS BY ASSETS

Number of bank holding companies, utilities, life insurance companies and financial service companies with 1982 assets of $1.9 billion or more.

Dallas **10**

Kansas City **5**

Seattle **6**

Indianapolis **4**

Milwaukee **4**

Portland **4**

Omaha **2**

Memphis **1**

Columbus, Ind. **0**

Source: Fortune and Forbes magazines

The Star/Robin Jefferies

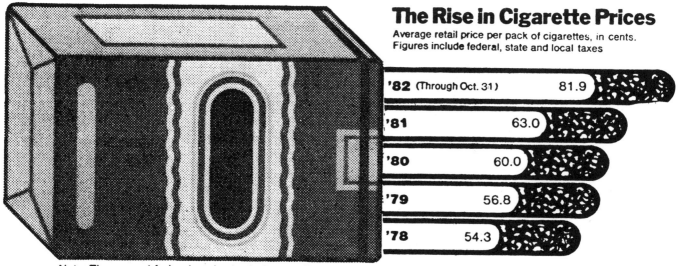

The Rise in Cigarette Prices

Average retail price per pack of cigarettes, in cents. Figures include federal, state and local taxes

'82 (Through Oct. 31) 81.9

'81 63.0

'80 60.0

'79 56.8

'78 54.3

Note: The current federal excise tax is 8 cents. The tax will increase to 16 cents in January.   Source: Tobacco Institute

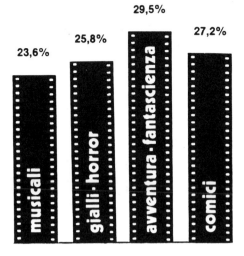

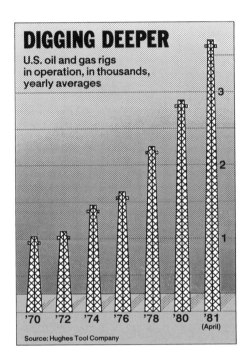

DIGGING DEEPER

U.S. oil and gas rigs
in operation, in thousands,
yearly averages

'70 '72 '74 '76 '78 '80 '81
(April)

Source: Hughes Tool Company

4. What Are Appropriate Uses?

The bar chart is most useful when you need to show individual numbers in a series or when you want to plot different items at the same time.

Time series. While the fever line gives an impression of the flow of a series of numbers, the bar gives more individual prominence to each amount being plotted, since each amount has a complete bar to itself, rather than being joined to all the others in the series. Thus it is appropriate to use this form of chart where the priority is to comprehend individual numbers within a series. If an overview of Dow Jones statistics for a few months is required, use a fever line. If a detailed look at the last two weeks' daily closings is what's needed, use a bar chart.

Non-time series. A completely different function of the bar chart is the comparison of different commodities plotted at one given time. Thus, for example, if you are showing the cost of renting an apartment in different cities across America, it would be appropriate to use a bar chart (and nonsensical to use a fever).

5. What Are Inappropriate Uses?

In some cases, there may simply be too many numbers to be able to show them all individually. The bars would then become too thin to make sense in terms of space. If you must place bars too close together their point as individual bars is lost. In a case where there is too much data, print that as a table or suggest that it be edited down. Only then should a bar chart be attempted.

As has been noted before, if the flow rather than a look at individual numbers is the prime consideration, then the bar chart is inappropriate.

6. What Are the Main Requirements?

Although bar charts can be achieved quickly and effectively by simple freehand drawing, they always look better when they have been ruled out neatly and accurately. They therefore take longer and consequently cost more to produce in man-hours than the simpler fever lines. Moreover, since there may be information about more than one commodity in the chart, there will be more labels. Thus typesetting costs will be greater, and the type will demand more careful planning—for instance, how it's placed within the design and how it's specified to make sure that it is readable. In turn, where more type is involved, more time should be allowed for proofreading and fact checking.

For a review of the materials and techniques to choose from, see page 35.

The illustrations of cigarettes, corporate headquarters, filmstrips, and oil wells show that almost anything which can be extended in one dimension can be used to depict itself in a chart.

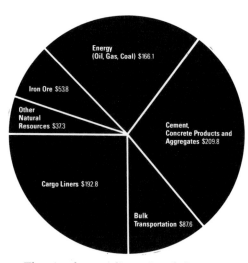

The simple two-dimensional figure shows lines radiating out from the central point to the circumference. The understated elegance of this breakdown of revenues is most effective.

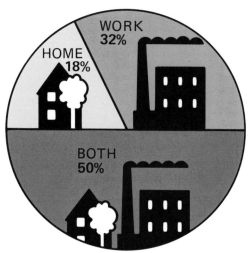

As an example of an illustrated pie chart, the use of computers by home and industry is shown by symbolic drawings inside the segments.

THE PIE CHART

If the names of the first two types of charts seem to be literal—the feverish fever chart, the blocklike bar chart—then the pie chart's obvious reflection of its source should come as no surprise. Slices of the pie, pieces of the cake, parts of the picture all describe the activities of this, the third type of graphic presentation.

Answering the six questions that follow will help you understand the characteristics particular to the pie chart.

1. What Is the Definition of a Pie Chart?

The pie chart, or divided circle, is a method of showing percentages of a whole. The complete circle, or whole pie, represents 100 percent. The "slices of the pie" are the divisions of that 100 percent, sized according to the magnitude of the percentage of each individual piece. For example, a 50 percent slice of the pie would be half of it.

2. What Are Its Main Elements?

In the basic form, there are merely two elements: the circle and the typically triangular shapes representing the subdivisions, or percentages, of the whole. The labels and the placing of them accurately in relation to their relevant segments are the only other elements needed to construct a pie.

3. What Variants Are There?

Pie charts are perhaps less flexible in their variations than fevers or bars, but nevertheless they can progress from the simple to the illustrated while still revealing their numbers efficiently.

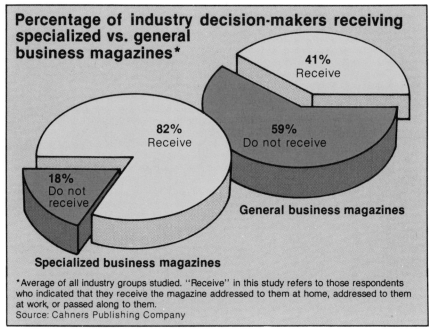

Using the third dimension allows the circle to be tilted back into an oval, giving the impression of it sitting on the page. Definition and emphasis can be put into the chart by separating the information and making almost a real "slice of the pie."

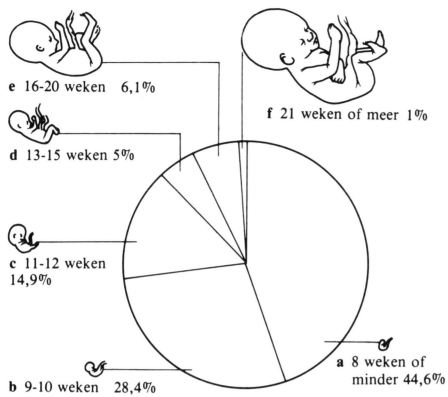

e 16-20 weken 6,1%

f 21 weken of meer 1%

d 13-15 weken 5%

c 11-12 weken
14,9%

b 9-10 weken 28,4%

a 8 weken of
minder 44,6%

The simplest of pie charts can be made visually more enlightening by the addition of line drawings, in this case showing the growth of the fetus. Note that in this example all the labeling is contained outside the pie segments, compared with all the examples on the opposite page.

A third dimension may be added to give the pie the impression of being solid, in the same way that the fever and bar can be treated. In this example, where it was possible, the percentages have been placed within the segments. The change in the total size of the labor force between 1979 and 1995 is reflected in the overall size of the circle.

More Workers Will Be in Their Prime

16-24 years old
Nonwhite   White

**25-54
years old**

**55 and
older**

20.9%   13.9%

3.7%

61.5%

3.2%   14.0%   10.9%

71.9%

LABOR FORCE: 102.9 MILLION

LABOR FORCE: 127.5 MILLION

1979

1995

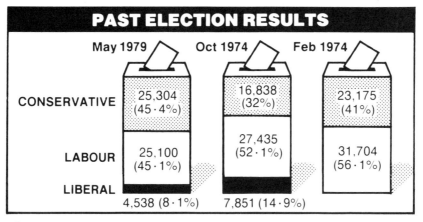

PAST ELECTION RESULTS

	May 1979	Oct 1974	Feb 1974
CONSERVATIVE	25,304 (45·4%)	16,838 (32%)	23,175 (41%)
LABOUR	25,100 (45·1%)	27,435 (52·1%)	31,704 (56·1%)
LIBERAL	4,538 (8·1%)	7,851 (14·9%)	

The noncircular pie chart could in some cases be seen as an example of a bar chart. It is included here, however, since it falls within the definition of a pie—a division of a shape into percentages. This example adds a little dimension and symbolism, thus showing the breakdown of voting patterns in England very clearly.

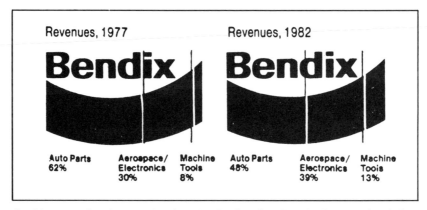

Revenues, 1977
Revenues, 1982

| Auto Parts 62% | Aerospace/ Electronics 30% | Machine Tools 8% | Auto Parts 48% | Aerospace/ Electronics 39% | Machine Tools 13% |

In this example of a noncircular pie the familiar Bendix logo is divided up to show revenues in 1977 and again in 1982.

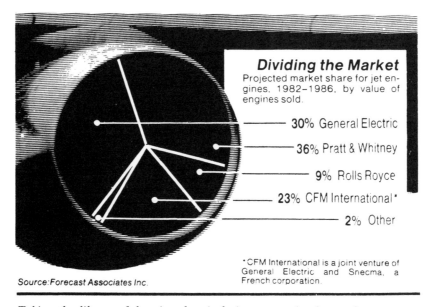

Dividing the Market

Projected market share for jet engines, 1982–1986, by value of engines sold.

- 30% General Electric
- 36% Pratt & Whitney
- 9% Rolls Royce
- 23% CFM International*
- 2% Other

*CFM International is a joint venture of General Electric and Snecma, a French corporation.

Source: Forecast Associates Inc.

Taking the liberty of drawing the circle in perspective in previous examples, it could now become the top of a bottle, the wheel of a car, or another pictorial symbol appropriate to the subject of the chart. Set into the jet engine of an aircraft, this pie is simply achieved, but makes the point immediately.

PORTRAIT OF A RECESSION
Percent share of the unemployed in July 1980

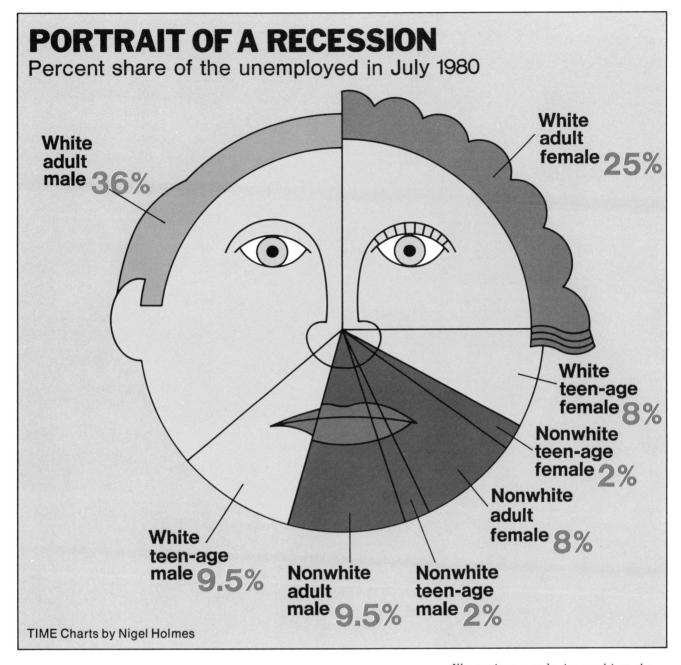

White adult male 36%

White adult female 25%

White teen-age female 8%

Nonwhite teen-age female 2%

Nonwhite adult female 8%

White teen-age male 9.5%

Nonwhite adult male 9.5%

Nonwhite teen-age male 2%

TIME Charts by Nigel Holmes

Illustrations may be inserted into the segments either to make up a whole (for instance, a face) or simply to show the content of each segment. The "portrait of recession" colors in the parts of the face in a way that is appropriate to the data being presented.

51

A variation of the pie as a device to record percentages of a whole is shown here with the circle representing a year with the months marked around it. This can then be used to record activities or other data for the year. Here the length and timing of the seasons are recorded. Dark gray is winter, light gray is spring or autumn, and white the summer.

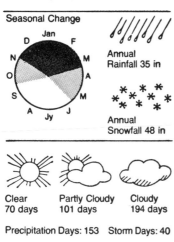

Seasonal Change

Jan
D · F
N · M
O · A
S · M
A · M
Jy · J

Annual
Rainfall 35 in

Annual
Snowfall 48 in

Clear
70 days

Partly Cloudy
101 days

Cloudy
194 days

Precipitation Days: 153   Storm Days: 40

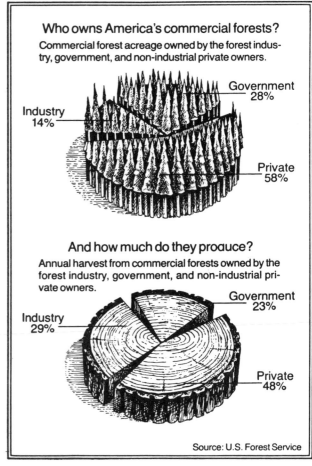

Who owns America's commercial forests?
Commercial forest acreage owned by the forest industry, government, and non-industrial private owners.

Industry
14%

Government
28%

Private
58%

And how much do they produce?
Annual harvest from commercial forests owned by the forest industry, government, and non-industrial private owners.

Industry
29%

Government
23%

Private
48%

Source: U.S. Forest Service

For a chart about tree growing, what could be more appropriate than a section through the tree itself! This image has the added advantage of being a known object (unlike the jet engine, which seldom has a chart inside it!), and the divisions are neatly made into axe cuts, or splits in the wood, emphasizing the "truth" of the picture.

WHERE IRAN'S OIL GOES

Percentage
of total
exports
(1st quarter
1980 estimates)

JAPAN 39%

W. GERMANY 21

BRITAIN 5
SPAIN 5
Other W. EUROPE 13
E. EUROPE 8
Others 9

In this example of a noncircular pie chart, Iran's oil output is symbolized by a barrel made into the face of a leader who had not yet outraged the world with his actions over American hostages.

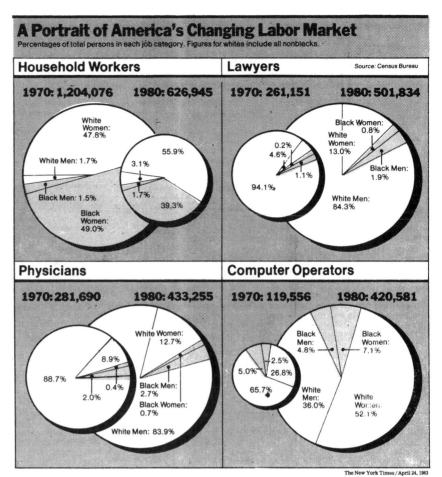

The New York Times / April 24, 1983

The labor market is the subject of this group of charts that show in the most obvious way the distinction between black and white workers. The more complex data here has made the labeling somewhat harder to follow, but a little study reveals much information.

4. What Are Appropriate Uses?

Budget breakdowns, share of the market, income and spending figures are all good uses of the pie chart. However, no more than eight segments of the pie should be attempted.

5. What Are Inappropriate Uses?

If there are more than eight divisions of the whole the slices become too small to be visually separate. If a slice is under 2 percent it will be very tiny, and it is justified only if it is a single statistic among larger amounts. Then the point will be visually compelling. It will, however, be hard to label and could lead to inconsistency. If there is too much to be said about each segment, meaning that the designer has to put sentences of type outside the pie, the chart should be reviewed: either edit the copy or consider the pie chart an inappropriate form.

In a noncircular pie chart—for instance, a map of the United States divided nongeographically to show, say, the ages of the population—make sure that the outline map is greatly simplified, or you run the risk of leaving an unclear impression of the figures involved. Too complicated a shape—one that departs too much from a circle—should be avoided as an image.

6. What Are the Main Requirements?

Devices can be bought that divide circles into percentages. Obviously they will save you a great deal of time and thus are an essential part of your equipment. Careful attention must be paid to the design of the pie chart, especially the planning of the type. Where possible this should go inside the segment it labels, but in cases where this is not possible a logical system for the position of the labels should be devised. As with most graphics: the simpler the better.

For a review of the materials and techniques available to choose from, see the appropriate discussion under the fever chart.

THE TABLE

Tabulation of information is the oldest form of graphic presentation. Indeed all the information that William Playfair used and charted was, before his invention of the fever, bar, and pie charts, printed in the form of tables.

In fact, a table is not a visualization of statistics at all. The fever, bar, and pie apply a graphic form to the statistics. They change one number into a shape that can be visually compared with another. The table, however, presents the number as a number, plain and simple. There will be many occasions when the designer will need this method of presentation, rather than the three already dealt with in this chapter.

To understand the special characteristics of the table, study the six questions that follow.

1. What Is the Definition of a Table?

An arrangement of information, often numbers, into columns or other organized ranks or groups is the basic feature of a table. There is no attempt to represent any of the numbers by a pictorial form.

2. What Are Its Main Elements?

The numbers, statistics, or facts themselves, with a heading to define or explain them, are the main part of any table. These numbers may be all that is necessary in a simple example, but as the information to be put across becomes more complex, a second element may have to be used: that is, a grid of lines or framework into which the numbers can be placed. At its simplest, this device helps the eye follow one column of figures across a table containing many columns. It helps one see which figures are related to each other and where a new group starts.

A secondary use of the grid is also to introduce a pictorial element into the table. Thus, the parallel bars of a prison cell door may be used as the grid to contain statistics about prison populations. The door is the frame and the internal grid for the table, and its presence also gives the reader an immediate clue as to the subject of the information about to be read.

3. What Variants Are There?

The table is responsible for mopping up many otherwise unclassifiable forms of graphic information. A typeset table of columns with headings and numbers is a table, of course, but so is a family tree, a calendar, or the index to a book. Shown here are the main variants.

Occasionally, symbols may be able to translate words into a truly international language. This table about an Italian diet food neatly substitutes drawings for words at each meal.

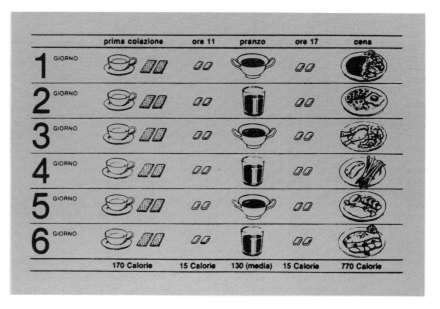

THE CARIBBEAN

MAGIC ISLE Haiti	PUNTA CANA Dominican Republic	CARAVELLE Guadeloupe	FORT ROYAL Guadeloupe	BUCCANEER'S CREEK Martinique
18	16	22	24	20

(A matrix of pictorial symbols indicating activities available at each resort. Annotations reading "PIGEON" appear in the FORT ROYAL column at two rows.)

A matrix of boxes with a vertical column of activities on the left and a horizontal list of locations across the top can be called a "what's available" table. The boxes can be filled in, or left blank, to indicate the occurrence of an activity (or the lack of it) at a particular location. In this example, symbols have been used to great effect to enliven a large table about holiday resorts. Although all that is needed to indicate the occurrence of an activity is a simple dot or check mark in the box, the designer's use of pictorial symbols instead nicely helps the eye and the mind distinguish different activities.

% who have confidence in Reagan's ability to:	Jan.	May	Sept.
Handle the economy	80%	82%	77%
Manage foreign policy	70%	76%	70%
Provide leadership for the country	81%	85%	80%
Appoint the best people to office	71%	76%	73%
Deal with organized labor	N.A.	N.A.	68%

Simple arrangements may be set into an illustrated border or frame. The report card with its corner turned up to give it a physical presence on the printed page and the jar of jelly beans to suggest the presidential subject of the report are merely graphic elements designed to contain the numbers.

SICK SYSTEMS total passengers in millions/adult cash fare

	1950	1955	1960	1965	1970	1975	1980
New York†	2,334/10¢	1,778/15¢	1,782/15¢	1,830/15¢	1,667/30¢	1,469/35¢	1,327/60¢
Chicago	1,325/15¢	990/20¢	847/25¢	792/25¢	661/45¢	650/45¢	692/80¢*
Philadelphia	828/12¢	622/18¢	568/22¢	419/25¢	274/30¢	314/35¢	330/65¢
Boston (subway; bus)	494▼/15;10¢	288/20;15¢	256/20;15¢	265/20;10¢	254/25;20¢	245/25;20¢	280/50;25¢
Detroit	332/15¢	216/20¢	130/25¢	112/25¢	112/40¢	77/45¢	70/60¢
Los Angeles	321/10¢	187/17¢	167/15¢	193/25¢	199/30¢	218/25¢	353/65¢
Baltimore	288/15¢	168/18¢	125/25¢	116/25¢	78/30¢	127/30¢	115/50¢
Pittsburgh	272/15¢	137/20¢	94/27¢	104/30¢	104/35¢	110/40¢	106/75¢
San Francisco	250/10¢	200/15¢	202/15¢	198/15¢	201/25¢	170/25¢	295/50¢
Atlanta	N.A./10¢	91/15¢	74/20¢	71/25¢	65/35¢	75/15¢	119/50¢
U.S. TOTALS	17,246	11,529	9,395	8,253	7,332	6,972	8,228

†Does not include transfers
*January 1, 1981
▼1951 figure

Source: American Public Transit Association
and local transit systems
TIME Chart by Nigel Holmes

(Above) The box containing the table may become an illustration in itself. For a table comparing the costs of transportation in various cities around the United States, the whole table has been turned into a vehicle, slanted at the same angle as that of the italic type inside the boxes to accentuate the feeling of movement.

(Below) The whole table may be set over a photograph, in this case one which gives a generalized idea about the "future" rather than a specific one about the subject. A number of tables and charts were used in this booklet about computers and the future of the office; photographs of space, galaxies, and the planets provided different but consistent backgrounds.

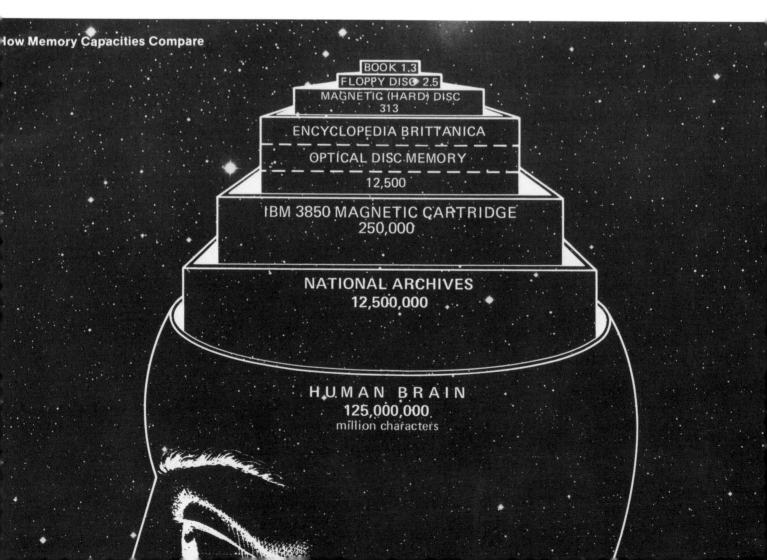

How Memory Capacities Compare

BOOK 1.3
FLOPPY DISC 2.5
MAGNETIC (HARD) DISC 313
ENCYCLOPEDIA BRITTANICA
OPTICAL DISC MEMORY
12,500
IBM 3850 MAGNETIC CARTRIDGE
250,000
NATIONAL ARCHIVES
12,500,000
HUMAN BRAIN
125,000,000
million characters

As an organization of information into an understandable form, the family tree is a further variation of the table. Many forms exist; this one uses a number of three-dimensional boxes in increasing sizes to emphasize the passage of time, and a heavy line connects the direct line of succession. It was published before Prince William had been named.

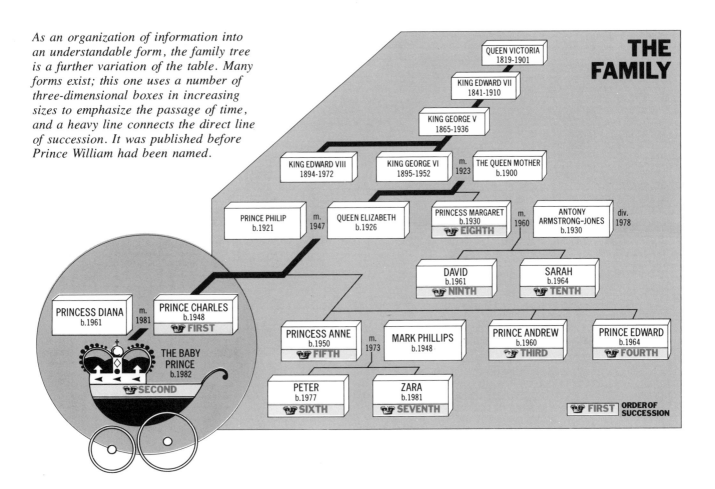

THE FAMILY

QUEEN VICTORIA
1819-1901

KING EDWARD VII
1841-1910

KING GEORGE V
1865-1936

KING EDWARD VIII
1894-1972

KING GEORGE VI
1895-1952

m. 1923

THE QUEEN MOTHER
b.1900

PRINCE PHILIP
b.1921

m. 1947

QUEEN ELIZABETH
b.1926

PRINCESS MARGARET
b.1930
EIGHTH

m. 1960

ANTONY ARMSTRONG-JONES
b.1930

div. 1978

DAVID
b.1961
NINTH

SARAH
b.1964
TENTH

PRINCESS DIANA
b.1961

m. 1981

PRINCE CHARLES
b.1948
FIRST

THE BABY PRINCE
b.1982
SECOND

PRINCESS ANNE
b.1950
FIFTH

m. 1973

MARK PHILLIPS
b.1948

PRINCE ANDREW
b.1960
THIRD

PRINCE EDWARD
b.1964
FOURTH

PETER
b.1977
SIXTH

ZARA
b.1981
SEVENTH

FIRST ORDER OF SUCCESSION

Here a symbol or symbolic illustration can be effectively used alongside each separate entry in a table.

WHAT DO YOU DO? Every day, or almost every day, the percent of people who:		EXERCISE OR JOG	35%	
		SPEND AN EVENING JUST TALKING TO SOMEONE	30%	
WATCH TELEVISION	72%	READ A BOOK	24%	
READ A NEWSPAPER	70%	PURSUE A HOBBY	23%	
LISTEN TO MUSIC AT HOME	46%	WORK IN THE GARDEN	22%	
TALK ON PHONE TO FRIENDS OR RELATIVES	45%	ENGAGE IN SEXUAL ACTIVITIES	11%	

57

A distance chart (mileage / kilometre table). Distances above the diagonal are in **kilometres**; distances below the diagonal are in **miles**.

	London	Aberdeen	Aberystwyth	Bath	Birmingham	Brighton	Bristol	Cambridge	Canterbury	Cardiff	Carlisle	Chester	Dover	Edinburgh	Exeter	Fishguard	Glasgow	Harwich	Holyhead	Hull	Inverness	Liverpool	Manchester	Newcastle upon Tyne	Norwich	Nottingham	Oxford	Penzance	Perth	Plymouth	Reading	Southampton	Stranraer	Stratford-upon-Avon	Swansea	York
London		869	379	186	187	85	190	96	98	248	491	301	123	648	272	421	643	128	427	344	898	336	318	448	184	205	90	466	717	344	64	125	662	152	309	334
Aberdeen	543		755	827	688	966	818	749	949	851	374	592	952	203	934	851	238	854	734	574	168	629	566	378	797	635	797	125	133	998	810	907	390	722	883	515
Aberystwyth	237	472		227	190	429	202	354	464	184	376	160	504	536	315	93	530	480	170	365	789	176	213	440	453	251	250	501	608	378	301	323	554	203	123	325
Bath	116	517	142		160	203	21	229	283	90	459	264	320	614	160	272	614	314	368	411	870	312	290	506	383	259	106	341	693	226	120	102	637	114	155	392
Birmingham	117	430	119	100		283	136	162	298	171	310	138	325	469	251	291	466	267	248	218	725	157	141	317	258	94	101	445	542	318	131	205	486	38	216	205
Brighton	53	604	268	127	177		242	190	118	298	584	424	125	747	275	467	742	218	520	438	998	434	410	544	277	299	181	467	816	342	115	101	762	237	350	429
Bristol	119	511	126	13	85	151		246	299	72	440	270	317	597	130	256	595	326	347	363	851	285	267	466	344	242	118	323	672	200	125	120	618	120	136	354
Cambridge	60	468	221	140	101	119	154		182	302	410	310	203	547	370	434	565	106	402	254	800	314	248	365	99	134	128	563	618	437	146	211	584	158	326	251
Canterbury	61	593	290	177	186	74	187	114		352	602	416	26	722	373	534	755	184	541	456	1011	450	429	557	251	304	194	552	834	440	174	210	778	274	416	448
Cardiff	155	532	115	56	107	186	45	189	220		472	243	373	629	192	181	629	381	344	394	885	320	301	498	400	272	174	386	706	262	176	195	650	158	66	386
Carlisle	307	234	235	290	194	365	275	256	376	295		230	629	157	555	486	155	520	360	248	413	202	189	93	454	296	418	749	232	626	461	536	174	360	512	187
Chester	188	370	100	165	86	265	169	194	260	152	144		448	379	400	254	381	430	155	219	637	29	67	285	352	149	245	581	462	459	288	352	405	176	259	178
Dover	77	595	315	200	203	78	198	127	16	233	393	280		750	394	560	784	210	562	458	1037	352	453	570	270	338	216	584	856	459	202	232	805	301	442	454
Edinburgh	405	127	335	384	293	467	373	342	451	393	98	237	469		714	637	72	656	523	371	254	360	349	174	592	437	579	907	70	781	611	693	202	509	669	312
Exeter	170	584	197	100	157	172	81	231	233	120	347	250	246	446		374	710	413	462	475	966	400	382	578	469	355	243	194	787	72	229	173	731	240	258	466
Fishguard	263	532	58	170	182	292	160	271	334	113	304	159	350	398	234		637	531	261	472	896	282	322	544	558	344	326	563	720	442	360	374	661	310	117	432
Glasgow	402	149	331	384	291	464	372	353	472	393	97	238	490	45	444	398		672	514	392	274	354	342	240	606	450	568	901	98	778	613	686	141	510	664	333
Harwich	80	534	300	196	167	136	204	66	115	238	325	269	131	410	258	332	420		512	360	910	416	360	474	104	246	208	595	726	474	195	261	696	264	445	360
Holyhead	267	459	106	230	155	325	217	251	338	215	225	97	351	327	289	163	321	320		347	770	170	200	419	501	282	339	654	592	530	381	443	534	277	323	307
Hull	215	359	228	257	136	274	227	159	285	246	155	137	286	232	297	295	245	225	217		640	202	155	194	245	147	301	672	442	546	350	405	414	267	466	61
Inverness	561	105	493	544	453	624	532	500	632	553	258	398	648	159	604	560	171	569	481	400		603	589	429	845	670	829	1155	187	1041	867	944	416	765	925	568
Liverpool	210	393	110	195	98	271	178	196	281	200	126	18	220	225	250	176	221	260	106	126	377		54	272	371	318	264	592	432	470	309	379	378	208	283	162
Manchester	199	354	133	181	88	256	167	155	268	188	118	42	283	218	239	201	214	225	125	97	368	34		226	293	114	246	573	418	450	288	358	362	186	344	114
Newcastle upon Tyne	280	236	275	316	198	340	291	228	348	311	58	178	356	109	361	340	150	296	262	121	268	170	141		413	259	416	779	254	656	448	518	262	362	552	133
Norwich	115	498	283	208	161	173	215	62	157	250	284	220	169	370	293	349	379	65	313	153	528	232	183	258		197	227	661	662	534	234	306	629	250	458	296
Nottingham	128	397	157	162	59	187	151	84	190	170	185	93	211	273	222	215	281	154	176	92	431	199	71	162	123		166	547	507	424	197	274	472	112	309	138
Oxford	56	498	156	66	63	113	74	80	121	109	261	153	135	362	152	204	355	130	212	188	518	165	154	260	142	104		435	650	309	42	104	594	62	227	296
Penzance	291	703	313	213	278	292	202	352	345	241	468	363	365	567	121	352	563	372	409	420	722	370	358	487	413	342	272		978	128	410	363	922	422	445	667
Perth	448	83	380	433	339	510	420	386	521	441	145	289	535	44	492	450	61	454	370	276	117	270	261	159	414	317	406	611		854	680	765	248	592	749	381
Plymouth	215	624	236	141	199	214	125	273	275	164	391	287	287	488	45	276	486	296	331	341	651	294	281	410	334	265	193	80	534		290	238	798	307	323	544
Reading	40	506	188	75	82	72	78	91	109	110	288	180	126	382	143	225	383	122	238	219	542	193	180	280	146	123	26	256	425	181		77	637	104	243	330
Southampton	78	567	202	64	128	63	75	132	131	122	335	220	145	433	108	234	429	163	277	253	590	237	224	324	191	171	65	227	478	149	48		715	171	258	125
Stranraer	414	244	346	398	304	476	386	365	486	406	109	253	503	126	457	413	88	435	334	259	260	236	226	164	393	295	371	576	155	499	398	447		536	688	662
Stratford-upon-Avon	95	451	127	71	24	148	75	99	171	99	225	110	188	318	150	194	319	165	173	167	478	130	116	226	156	70	39	264	370	192	65	107	335		197	246
Swansea	193	552	77	97	135	219	85	204	260	41	320	162	276	418	161	73	415	278	202	291	578	177	215	345	286	193	142	278	468	202	152	161	430	123		435
York	209	322	203	245	128	268	221	157	280	241	117	111	284	195	291	270	208	225	192	38	355	101	71	83	185	86	185	417	238	340	206	78	414	154	272	

A distance or mileage table is based on a matrix that indicates the distance between any of the places listed. In this elegant and compact example from England, the distance can be read in either miles or kilometers.

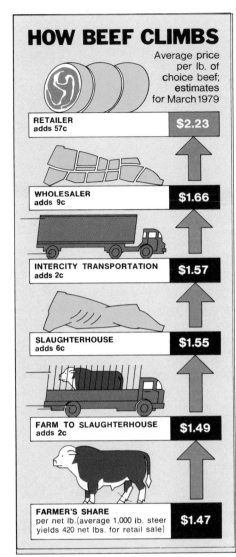

HOW BEEF CLIMBS

Average price per lb. of choice beef; estimates for March 1979

RETAILER adds 57c — **$2.23**

WHOLESALER adds 9c — **$1.66**

INTERCITY TRANSPORTATION adds 2c — **$1.57**

SLAUGHTERHOUSE adds 6c — **$1.55**

FARM TO SLAUGHTERHOUSE adds 2c — **$1.49**

FARMER'S SHARE per net lb. (average 1,000 lb. steer yields 420 net lbs. for retail sale) — **$1.47**

Where stages in the development of the subject of a table can be clearly identified and drawn, a further variant is to increase the amount of illustration so that it plays a role equal to the numbers. This obviously uses more space, but also helps the reader understand the meaning of the table. Here the increase in the price of a pound of beef is broken down into stages, with a little illustration to show the form of the meat at each one.

4. What Are Appropriate Uses?

There are at least six ways to use the table to best effect:

1. Where there are great differences between high and low numbers in a series that would otherwise be plotted, it is appropriate to set the figures down as a table rather than try to find a scale that will enable them to be charted.

2. Where there is a great amount of information, it is usually more space saving to set out the facts in type as a table.

3. Where different types of information are listed alongside one another, and thus where charting them alongside one another would be a pointless visual exercise, use a table.

4. Where it is necessary to read the exact numbers or quantities in a given set of statistics, the table is the only way to be sure that this is achieved. More than one decimal point, for instance, can never be seen on a chart, unless the scale is so minutely ruled out that it runs the risk of being confusing.

5. If after plotting out a multiple-line fever chart, the resulting shape is a tangle of criss-crossing and intersecting lines that looks, and therefore is, confusing, it is appropriate to tabulate the material for clarity.

6. Certain forms of graphic information just *are* tables. Calendars, timetables, duty rosters, family trees, frequency diagrams are what they are. There is no point in trying to force a different form onto the material. The designer's time and energy should be spent making them legible, elegant, and approachable.

5. What Are Inappropriate Uses?

When the statistics are so simple that they can be plotted out in a time series, you would be missing a graphic opportunity not to do so. The visual impact of the figures when charted, especially if those figures are themselves rounded off or generalized, is far greater than a dull listing of them in type.

6. What Are the Main Requirements?

The table communicates information only through the figures themselves, not through any other visual means. Therefore it is advisable always to present, even at a rough stage, the table in properly set or typewritten type, not as handwritten numbers. Since many typewriters have quite sophisticated tabulation functions, this may be achieved relatively easily, with a little effort and time. As for the finished art, the main cost element will of course be the typesetting. Since the nature of tables is to be able to handle large amounts of information, your type bill will certainly be higher than for any other type of chart.

While you consider various ways to achieve readable tables, review the discussion of materials and techniques under the fever chart earlier in this chapter.

Chapter Three
How to Approach and Analyze Your Assignments

THIS CHAPTER will discuss four steps you should take to analyze any assignment.

Step 1. Identify the reader/user of the chart: so that you know who your audience is.

Step 2. Select a production method: so that you can complete the job within the time allowed and according to the client's budget.

Step 3. Review the numbers: so that the final chart makes the desired point by displaying the right numerical information.

Step 4. Find the right symbol: so that the information is amplified by a visual presentation which helps tell the story.

STEP 1. IDENTIFY THE READER/USER OF THE CHART

The most important part of analyzing your assignment is having a clear understanding of who is going to see your finished work. The facts must come through strong and clear, of course, in whatever form they appear when charted, but that form must engage the reader's interest in the first place. If you can attract the viewer to look at the chart, then he or she will read it. But who is that person?

Who will read the chart? Is it going to a specialist? Is it, for example, a medical chart for medical students? A financial chart for accountants? A geological chart for geologists? Or perhaps the medical chart will be read by a patient, the financial one by a casual reader of a newspaper's business page, and the geological chart of the early ages of the earth by young school children.

In the course of finding out who your audience is, consider the following questions:

Are they educated and what is their lifestyle: scholarly, scientific, general management, the arts, students, the "general public"?

Are you talking to people about their own subject with which they are fully conversant and to which you are adding some new information, perhaps updating some facts and figures?

Are you informing them about something that is entirely new to them, on a subject they do not necessarily know?

Are you entertaining people who do not need the information but might be interested by it?

Are you educating people, as in a classroom?

Is there a political motive behind your chart? Perhaps you want to convince someone to take action as a result of seeing your revelations on paper or a slide.

Is the chart to be published or privately shown to a single individual on a one-to-one basis?

A basic consideration must be how serious your presentation will be. Setting aside common sense (do not make fun out of cancer statistics, for example, or the number of Catholics compared with Protestants in Northern Ireland), this can generally be discussed between artist and client. It does not necessarily follow that if you are addressing specialists in a given subject, the charts should be dry and serious, or conversely that the audience for a general interest magazine should be titillated by decoration, illustration, and other "art" tricks. Specialists in fact need as much coddling and persuasion to read as the general public does. They may not after all be specialists in reading charts, although they are extremely knowledgeable about the subject matter.

Should you be doing a presentation to someone about the high costs of, say, running a particular program in your office, do not annoy the viewer with a highly finished set of charts, which themselves look expensive. It would be too easy for the person to retort that he or she can see why it all costs so much.

In any case consider carefully the tone of your chart. That can only come from knowing who will read it.

Sometimes, however, decisions about the recipient of the chart may not be quite so important because you are contributing to a format that has already been worked out. Newspapers and magazines already have a style and in a sense that style is attuned to the readers they expect. They have already gone through the thinking about who is going to read the information.

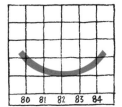

An upturn in the figures can be made to look happy too.

STEP 2. SELECT A PRODUCTION METHOD

If you have a choice in the matter, take some time to consider the best way of presenting your information. You may choose to draw it simply by hand, or make a more finished version (which still stops short of actual printing), or make a slide or any of the other methods outlined below.

The methods are divided into two categories and then subdivided within those broad fields. The first is one-time methods, the second publication methods.

One-Time Methods. There are a variety of one-time methods from which you can choose: by hand, by adding press type, with slides, through overhead projection, and by computer.

By Hand. Using broad magic markers, bold and colorful graphs can be made very quickly. Where you are presenting information on a one-to-one basis it might even be interesting to regard the chart as a piece of performance art, constructing it as you make the presentation in front of the client. Practice this beforehand so that the show goes smoothly. Work out the scale ahead of time and think about how the best effect will be achieved, and then proceed as though you were animating a cartoon, leaving the important figures to the end.

An example would be that after you say that "Sales declined in 1980, 1981, and 1982 but picked up again in 1983 and 1984, which has meant that in general an air of happiness has prevailed in our divi-

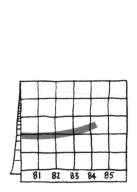
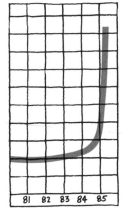

In the course of a personal presentation about sales, let the figures literally unfold before the viewers' eyes as you draw the line upward to its dramatic conclusion in 1985.

You can choose from a large variety of charting tapes to lend authority, clarity, and neatness to your artwork. Like the tone sheets described earlier, they are adhesive backed. Some have more flexibility than others when it comes to bending the line to follow a curve.

Pressure-sensitive lettering can be bought in sizes ranging from 6 point (very difficult to handle) to 6 inches high (good for signs or large presentations to a group of people).

sion," you then complete the presentation by drawing a circle around the line, making it into a mouth.

If you have some dramatic figures up your sleeve, fold over the paper you are working on to hide the fact that you can easily go off the top of the chart when you need to show a sudden rise (or drop). Then after you say "little progress was made in sales from 1982 to 1984," draw the line upwards as you unfold the paper and say "but they leapt up in 1985."

Adding Press Type. Taking longer, because of the degrees of finish attainable, is the method of drawing a chart by hand, but adding Letraset or another kind of graphic aid. There are many products that can help make large charts look more authoritative simply because they look "printed." They may also have the advantage of being more readable! If you are addressing a group of people at a conference or a press reception and need to make some points quickly as part of your speech, large cards can be designed for you to hold up in front of the group.

Charting tapes in all colors, widths, and patterns are available. They may be transparent or opaque. They are very easy to use and relatively inexpensive for the finish they impart to the work. And if you are working late in the office on December 24, they also do good duty for last minute gift wrapping.

Slides. When there is enough time to produce it, a slide presentation is very effective. It is obviously more formal, in that a group of people have to gather in one room at the same time, and a projector, screen, and darkening of the room all have to be coordinated. The advantages of enlargement are obvious in that many people can see and concentrate on the same information at one time. Certain companies specialize in graphic slides and generally handle all stages of production after the facts to be shown are discussed.

You will certainly have to prepare a rough for this discussion to hinge on, but remember that some basic principles should be adhered to when preparing slide presentations. Unlike print media where small details can be included in the design, slides must be simple or the audience will be lost. However, just as a magazine holds its readers through changes of pace, color, size of image, and concept, a slide show should be orchestrated with the same end in mind: to keep the audience watching.

Since slide images are viewed differently from printed images—in other words, the viewer has little or no control over the timing of the show—theories about which areas of a picture the eyes are drawn to first are more important considerations than those involving print, where they may wander at ease over an image until everything has been seen in the reader's own time. Theoretically people look first at the upper left-hand area of a projected slide, so use that space for important information. The eye naturally moves from left to right; therefore a sequence of events that must be read in order should obviously flow from left to right.

Do not use too many words. More than about 20 won't be read. If necessary, make color Xeroxes of the slides themselves, but at the least a typed summary of the facts and figures should be given to everyone to take away after the presentation.

Allow more space around the words than you would for print, and keep the typefaces simple. Do not make the contrast too great between image or type in relation to the background, unless you know that the slides will be shown in a room that is not properly darkened. In that case the higher contrast will make the image more visible.

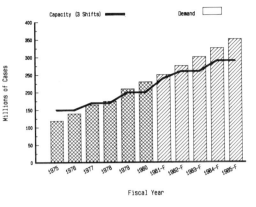

Overhead Projection. A method used during lectures or in a classroom is overhead projection. The system of mirrors, lights, and lenses casts an image—this may be a complete chart, for example—onto a screen from a flat original. The usefulness of overhead projection is that it allows the lecturer to make additions to a partly started image or to draw the whole chart, say, onto a prepared grid.

The ease with which a sequential set of statistics can be shown is a good reason for choosing this method over a simple slide presentation, where each of the images are complete when projected.

By placing your original art under a sheet of acetate, changes or additions can be made on top of the art with a grease pencil, and then cleaned off, perhaps to show a different line. Alternately, a new sheet of acetate can be used; of course this could already be prepared with material drawn on it. By this method a base grid may be shown, first with nothing on it; then with an acetate overlay, a fever line may be added to the grid; then another; and so on until the chart is complete.

Computer. At present most computers cannot produce a printout of sufficient quality for magazine or newspaper reproduction. Typically the computer printer draws with a rather shaky line that is unacceptable. This is because the output is a translation from one medium to another. In video, lines and bright saturated colors are made with light that flickers minutely, but that the eyes read as continuous. Where areas of a chart can be filled very easily with color on a TV screen, the graphic printer used to produce so-called hard copy of that image has only a set of four pens with which to draw thin lines. The pens must be made to cover the area by laboriously traversing the paper back and forth until the shape is filled in.

Most graphic printers, therefore, are useful for one-time presentations or very short runs only, since their output is more a reminder of what was on the screen than an end in itself.

However, there are graphic printers available that are excellent for a slightly different reason. The Sharp model is small, very easy to use, and produces bars, pies, and fevers in four colors on a roll of paper 2 inches wide. It is a sophisticated graphic printing calculator rather than a computer, and it can show an artist in a few minutes the "shape" of the statistics. This can be used as a basis for a rough sketch of the chart, or since it is very accurate, you can enlarge the small output and trace off the lines. As such this machine is a very good tool for the preliminary stages of chart making.

Sharp's more complicated model, the PC 1500, is a programmable

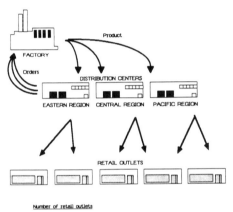

North American Distribution System

Number of retail outlets

EASTERN REGION:	120	independents
	35	chains
CENTRAL REGION:	85	independents
	32	chains
PACIFIC REGION:	89	independents
	16	chains

Printed on Apple's Dot Matrix Printer

(Above) Computers are becoming increasingly sophisticated at printing out figures, graphs, and diagrams. However, they are best used as tools to help the artist find graphic form for the statistics, which will be redrawn later using the printed output as a basis for the artwork.

(Right) Computers are now used extensively in the production of slides. The speed with which an image can be achieved is a great advantage where fast turnaround of information is required. A 35mm camera is built into the system.

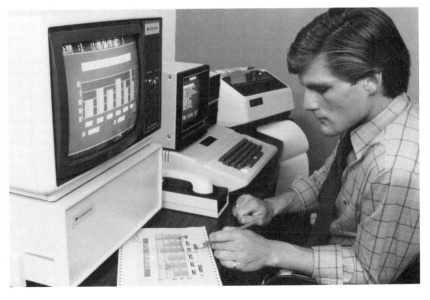

hand-held computer with an optional printer that produces much the same output, but it must be programmed to do all the things that the smaller and cheaper model has already built into it.

Until such time as computer companies can make more sophisticated printing machines that produce "hard copy" of a sufficient standard for publication, computers will serve for chart making as (1) a useful way to produce one, two, or possibly even 10 copies of a chart that it has plotted and (2) a tool for plotting the figures from which an artist can develop a chart based on, or traced from, the computer image.

If the art director is interested in achieving a computer "look" for a chart—for instance, if the subject of the chart is about computers—then you may choose to use the available printout, however wobbly the line may be. Computer printers have their own look or style, just as an illustrator's or artist's work has a recognizable style—that is part of what an art director is expecting when he or she assigns a job. Part of the trouble with the computer's *drawing* style is that it is not quite suited to chart making. The precise lines required for most graphic representations are more easily achieved by a human hand. So the key to understanding how a computer can help in this field is to use its *computing* powers, rather than its poor ability to mimic drawing.

The accompanying sequence of illustrations shows what is possible in terms of looking at the same set of figures from different points of

view. (A) shows the original hand-plotted and -drawn version. The same data was fed into a computer, which then produced all the other versions (B through F), turning the image around to see what it looks like from a variety of angles, ending up with a view from the side opposite the original. This offers an artist multiple choices for the final chart, for one of these computer printouts can be used as a base from which to trace clear, straight lines.

In sum, there are a few occasions when the computer's own drawing may be quite the correct type of illustration, but more generally—and much more usefully—it should be thought of as the metallic friend whose whizzes, whirrs, and beeps will remove the drudgery of plotting the movements of the Dow Jones Index more quickly and with fewer complaints than you could!

The Computer at Daily Newspapers. The computer is used daily at *The New York Times*, for instance, and has resulted in a 15-to-20-hour-per-week saving of time taken to produce roughly 74 charts and 48 maps. The aim is also to integrate the computer output with the rest of the newspaper page: type, photos, and rules.

In the next part of this chapter, there will be a discussion of a new breed of "production" computers which can make lines that the chart artist would be proud to have drawn. Their drawback: They cost a million dollars.

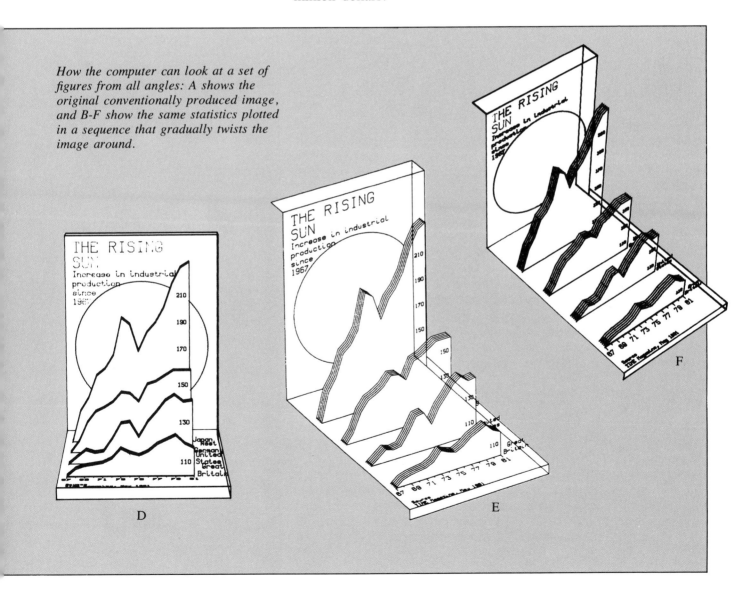

How the computer can look at a set of figures from all angles: A shows the original conventionally produced image, and B-F show the same statistics plotted in a sequence that gradually twists the image around.

Publication Methods. The three methods available for publication include flat art, flat art with multiple overlays for color, and computer applications.

Flat Art. Work for publication differs from the one-time category in an important respect: The viewer does not see what you produce; it has to go through another process—generally, a printer.

The best way to produce flat art is to work out the lines of the graph completely on tracing paper and then place that under a sheet of Mylar or other drafting film. This is transparent and thus enables you to trace your worked-out shapes onto it clearly and without changes. The aim is to produce as neat a piece of work as possible, so that the platemaker (and therefore ultimately the printer) can do the best possible job of getting your work onto the printed page.

There is no reason for artwork to be sloppy. Do not equate doing a job quickly with doing it messily. Give the printer no chance to blame you for a poorly printed final job.

When working in black and white remember that, photographically, red film comes out black, so covering a large area of black within, say, a box, can be easily achieved by putting down red film cut to the size of the box, rather than by laboriously painting up to the edges of the rules. This can be put down directly on the original Mylar drawing and need not be on a separate overlay. However, if you are using Letratone (or a similar product) to give the chart variations in black tones, it is a good idea to put those areas on a separate sheet, so as to avoid harming the lines of the original drawing with the tiny cut marks around the tonal areas bounded by lines. You must use a light box for this procedure, especially when dealing with the darker tones.

The platemaker/printer will combine the two "flaps," or overlays, into one black image.

Where a photo is being used as the background to a chart, mount it, or a copy, to the correct size onto a board and then proceed as with an ordinary chart. Remember to specify to the printer that the photo is to be treated as a halftone, while the chart overprints, or drops out to white, in line. Take care that the numbers are readable against the photographic background. If you are not sure whether they should be black or white numbers (perhaps because the background is in the 50 percent range: not very light, suggesting black numbers, and not very dark, suggesting white ones), then put them inside black or white boxes so that there can be no doubt about their readability. The graph line itself will generally present no problem of this nature as the eye will naturally follow it across an area of darkness or lightness on the background and fill in the gaps unconsciously.

Cover your artwork with a protective sheet of tracing paper and add any notes to that to clarify your intentions to the printer. Put your phone number on it too, so that you can be contacted if there are any questions.

A good relationship with the printer cannot be stressed strongly enough. It is sometimes difficult to write down clearly an instruction when a short conversation is much more efficient. Too often artists and designers complain that "the printer did a bad job." Since the printer is the person who actually produces your work in sufficient quantity for people to see it, talk to him or her first before complaining about it after it's too late. If you genuinely express an interest in what happens to the work after it has left your hands, you will be pleasantly surprised by the desires of printers to do a really good job of printing your work. After all, it is just as much an advertisement for their work as yours.

Flat Art with Multiple Overlays for Color. Charts look their best when the color used is bright and "clean." In general using primary colors is

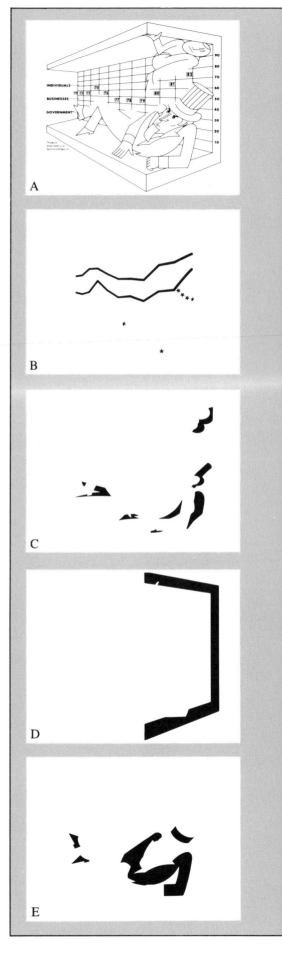

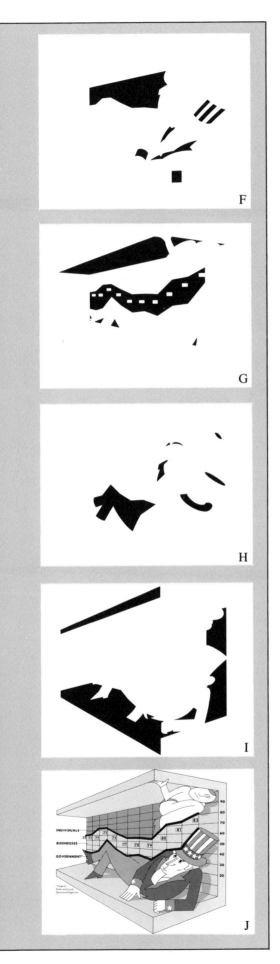

preferable to using secondary colors. The color is not being used to make an artistic impression; rather it is to distinguish between one object or set of numbers and others. Therefore the clearest, most obvious colors should always be your first choice. Clean colors are those that are made up of only two of the four process, or printing, colors. These will always appear the brightest. More subtle tones using all the process colors are useful for background tints—to make the bright colors of the bars or fever lines stand out in front.

The best results for a bright, clean look are always obtained by pre-separating the color for the printer to lay in the tints you require. These separations are cut by hand from "Amberlith" or "Rubylith," which is a very thin layer of amber- or ruby-colored film "stuck" to a clear sheet of acetate. After your Mylar drawing of the chart is complete, a sheet of this product is put on top of the artwork, taped onto the drawing board and keyed into position with register marks, and then cut with a scalpel or X-acto knife. Those parts that will not print are cut away from the layer of amber film exposing the clear "carrier" acetate underneath. The Amberlith is itself transparent, but represents black to the platemaker's camera. The printer then lays in tints of the percentage specified by you at the correct screen size for the particular publication in which the work will appear. To avoid any mistakes a separate sheet of Amberlith should be used for each color or tint of a color, but this may result in too many having to be cut. As long as the printer is aware of the situation, where colors or tints of colors do not abut, they can be cut from the same sheet. (Where they do abut, separate overlays must be prepared.)

This method may seem long and involved, and it does need careful marking up on the part of the artist, who must also know what percentages of the four process colors make up the particular tints wanted. But by using books of tints supplied by most good printers, the artist will find that the subtlety of the process is considerable and that far more colors can be obtained than might have been imagined from his or her own mind's palette. However, despite all this reliance on technicalities and considering not being able to see the finished art in color before it goes to press, the final printed result will always be crisper and brighter than if a full-color painting or colored drawing had been used as the original artwork, which then had to be separated photographically by the printer. What is more, the brighter, cleaner look achieved with preseparated art is in keeping with the nature of chart making, where information should generally take precedence over the artful look of the piece. Thus in the use of preseparated art the form (crisp and precise) follows the content (the exactness of numbers and statistics).

Here is an example of the complete artwork for one job. Either drawn onto Mylar, or cut from Amberlith by the artist, all these overlays were necessary for the production of the chart that is shown in color on page 121. In the list below, Y = yellow; R = red (magenta); B = blue (cyan); K = black (key). All the numbers are percentages of those four process colors.

A. 100K (the keyline, base art)

B. Stars: white
   Top line: 20R, 80B
   Bottom line: 100Y, 100R

C. 30K

D. 100Y, 40R

E. 20R, 80B

F. 100Y, 100R

G. 100Y, 60B

H. 20K

I. 100Y, 10R

J. The final piece, shown here in black and white. Certain parts of the chart – the flesh, for example – did not have an overlay, but were marked up for the printers to fill in themselves.

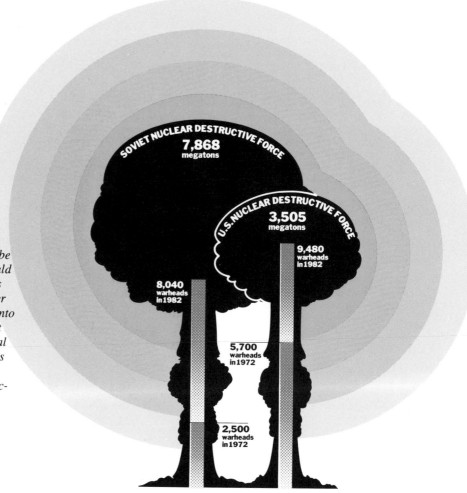

Subtle gradations of color can easily be achieved by computer. While one would normally specify differences of no less than 10% between tones for the printer to translate your Amberlith overlays into screens, the computer allows an artist to specify more subtle changes in tonal value. So-called airbrushed gradations in tone from dark to light can also be programmed into the computer, replacing conventional camera-ready airbrushing or gradated-screen adhesive film. In its original magazine use the story ran on top of the very pale computer-generated tints around the mushroom clouds of this bar chart.

It takes no longer to cut amber overlays representing tints of colors for the printer to lay in than it does to paint those colors or to cut colored film overlays and stick them down onto a drawing. The difference in the printed result, however, is immense.

When delivering a preseparated chart, include a colored pencil overlay on tracing paper to give both the client and the printer an idea of the colors you want. Do not assume that many clients can visualize what a face that looks like bright amber from the Amberlith overlay will actually look like when printed in the 20 percent yellow and 10 percent red that the overlay represents.

You may also be surprised at how few graphics people know the system of preseparating. You will have to explain that the method you have used is going to give him or her the best results. Suggest a meeting with the production staff if there is one or if not, with the printer so that everyone understands the technical side of the job. Until trust is built up among artist, art director, client, and printer you may find yourself doing fairly elaborate colored pencil sketches to accompany the final art. Leave enough time for this part of the job when you are planning how long it all will take.

Computer Applications. Although massively expensive in themselves, computers can be used for certain types of work in this field. Their use will undoubtedly increase as more platemakers, separators, and printers understand the quality improvement and ultimate cost-saving effects of using them.

The primary use of this new breed of computer is to produce separations. A transparency of a piece of flat art is scanned into the machine and the image appears on the TV screen at the operator's console. By using various buttons and commands from a typewriter-like keyboard the operator can retouch the image in an amazingly diverse number of ways. Colors can be changed, whole backgrounds laid in, spots removed, and an image changed if necessary. It is in its capacity as a coloring machine that it is useful to the chartmaker. By scanning just the Mylar drawing, the computer can fill in the colors that would otherwise have been cut from the Amberlith. It then produces the actual film that the plates will be made from, so that a whole stage in the production process is eliminated. What's more the computer "cuts" overlays very much more accurately than a hand. The resulting image fits perfectly together, with colors abutting one another exactly. Holding lines can even be eliminated at the touch of a button.

Some of these new machines have additional magic built into them.

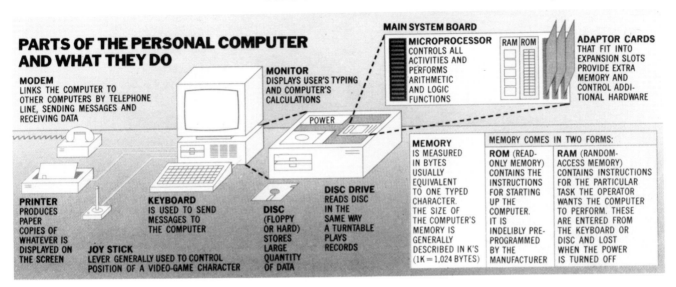

PARTS OF THE PERSONAL COMPUTER AND WHAT THEY DO

MODEM
LINKS THE COMPUTER TO OTHER COMPUTERS BY TELEPHONE LINE, SENDING MESSAGES AND RECEIVING DATA

MONITOR
DISPLAYS USER'S TYPING AND COMPUTER'S CALCULATIONS

MAIN SYSTEM BOARD

MICROPROCESSOR CONTROLS ALL ACTIVITIES AND PERFORMS ARITHMETIC AND LOGIC FUNCTIONS

RAM ROM

ADAPTOR CARDS THAT FIT INTO EXPANSION SLOTS PROVIDE EXTRA MEMORY AND CONTROL ADDITIONAL HARDWARE

POWER

PRINTER PRODUCES PAPER COPIES OF WHATEVER IS DISPLAYED ON THE SCREEN

JOY STICK LEVER GENERALLY USED TO CONTROL POSITION OF A VIDEO-GAME CHARACTER

KEYBOARD IS USED TO SEND MESSAGES TO THE COMPUTER

DISC (FLOPPY OR HARD) STORES LARGE QUANTITY OF DATA

DISC DRIVE READS DISC IN THE SAME WAY A TURNTABLE PLAYS RECORDS

MEMORY IS MEASURED IN BYTES USUALLY EQUIVALENT TO ONE TYPED CHARACTER. THE SIZE OF THE COMPUTER'S MEMORY IS GENERALLY DESCRIBED IN K'S (1K = 1,024 BYTES)

MEMORY COMES IN TWO FORMS:

ROM (READ-ONLY MEMORY) CONTAINS THE INSTRUCTIONS FOR STARTING UP THE COMPUTER. IT IS INDELIBLY PRE-PROGRAMMED BY THE MANUFACTURER

RAM (RANDOM-ACCESS MEMORY) CONTAINS INSTRUCTIONS FOR THE PARTICULAR TASK THE OPERATOR WANTS THE COMPUTER TO PERFORM. THESE ARE ENTERED FROM THE KEYBOARD OR DISC AND LOST WHEN THE POWER IS TURNED OFF

Originally in full color, this table of parts was "drawn" on a computer, which has the capacity to generate straight lines, circles, and ellipses. Printed as part of a lengthy magazine feature about computers, this in fact took longer to create on one than if it had been drawn by hand. But as an exercise in getting a machine to draw exactly what you want, it worked.

By using the airbrush mode the artist can specify graduations of color that would otherwise have to be airbrushed—a process taking up a lot of time and requiring great skill. The computing time is greater for these "vignettes" than for normal "painting by numbers" and thus more expensive (and the cost of producing art this way is still considerable), but the result is exciting and can already be seen starting to appear in magazines and advertisements. It is worth inquiring at the outset of a job whether the printer or platemaker has access to a computer such as this; more and more will have them. It will save you time and simultaneously allow you to do much more ambitious things.

The accompanying diagram was entirely drawn by such a computer. The lines were constructed and the resulting shape filled in with color all by pressing buttons and using a light pen. The process took about seven hours from start to final film.

STEP 3. REVIEW THE NUMBERS

The information from which you will construct a chart or graph will come to you in many different ways. The numbers involved may be prepared by the person requesting the chart, or they may possibly be in the form of a chart itself, which you are asked to redesign. You may have taken them down over the phone or noted them at a meeting. They may be typed, scribbled on a piece of paper, or hidden in a table of figures, a government report, or a press release. Whatever the case they will need to be organized and certain questions asked.

What Are the Key Relationships among the Numbers?

It is crucial to decide what the chart is trying to say—what point is being made. It may be better to show the beginning and end of a series of numbers, and no more. The intermediate numbers may be interesting but irrelevant to the story and may dilute the impact of the "it goes from here to here, in such and such a time" aspect.

Of course you must plot all the figures out for yourself to make sure that by just taking the beginning and end of a series you are not hiding some ups and downs that may be very important to the story. You will probably show this plot to your boss, art director, or client to discuss which parts are important. It may also be of value to see what the "shape" of the line is, for this can suggest a direction for the whole image and illustration, if it is to be that sort of chart. Finally, it will help you decide which type of graph is the best for the job; should it be a bar, a fever, or another type of chart?

Remember that your client may not have a clear idea of what the key relationships are, so you must study this first plot carefully and try to extrapolate some meaning from it yourself. Every chart tells a story. Or should.

What Are the Second Ranking Relationships?
Having decided what is important—what the eye is going to be drawn to first—consider the rest. How will you relegate the other numbers to the background but still make them readable to the enthusiast who wants the whole after the initial graphic blast?

How Do You Select the Right Horizontal and Vertical Scales?
If you are working within a space given by an art director these decisions will be limited by that space, but if there are no limitations, or if you must set your own, you can begin by simply plotting out the numbers very roughly and without measuring accurately to see what sort of space is actually going to be taken up. Once you have this rough plot you can judge impartially whether or not it represents the data in raw form. Start at least by numbering the scale from 0 at the base line and mark off points on the vertical scale at about half-inch intervals up to a height of about 6 inches. Do the same along the time axis (baseline). You will end up with a square that has equal boxes. Of course there may be more years than you have marked lines for, and, similarly for the vertical scale, but you will at least have a comfortably sized working tool on which to start your plotting and from which you can expand upwards or sideways. If you are intent on showing the steep increase in the price of gas, for instance, you will probably want to choose a scale in which the horizontal grid lines are further apart than the vertical grid lines, so that the resulting line is steep. Or if you want to show the long, slow recovery of the building industry the reverse may suit your purposes. If you have facts for a great many years, and each year's number is important, you may be forced into a long, thin rectangle simply to accommodate all those years.

None of these choices of scale should be considered distortions. It is simply a case of finding the most practical and clear way of presenting the facts. Nor should it be considered distortion to break a scale: that is, to start from somewhere other than 0. In some cases it will be a choice between doing this or not do a chart at all. Should you be asked to make comparisons within a set of numbers that includes some in units of tens and some in thousands, then the scale must break in the middle. As long as it is graphically clear, it is the only practical solution. Logarithmic scales can take care of some of the problems involving huge differences in numbers, but these complicated formulas are not recommended for everyday use, since they almost always need a footnote to be understood by the nonmathematical reader.

If it does become impossible to chart the salient information because of large swings in the amounts involved, then it would be better to set out the data in tabular form and not try to find the "appropriate" scale.

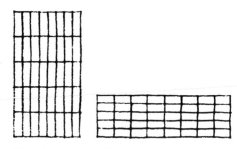

The choice of grid can accentuate the shape of the numbers. As long as it does not distort the information, the upright grid (left) will show a steeper climb, or drop, in a set of figures. Likewise a long, slow increase or decrease will best be served by a horizontal shape (right).

In summary: Look at the figures, plot them out roughly, ascertain whether or not they represent the story as you see it, change the scale if necessary to accommodate or to highlight a group of closely clustered statistics, and decide on a scale that can be comfortably assimilated in one look. Exaggeration by elongation or widening is not distortion until it actually does distort the information. To a certain extent choosing the scale is done by trial and error; the key is recognizing the right answer during that trial period.

What Is the Context?

Figures by themselves cannot be perceived as large or small until they are compared with others. Unemployment at 5 percent may be terrible if it was 2.5 percent the year before. Twice as many people are out of a job. On the other hand, if it has come down from 10 percent for the previous year, then things are looking up for those who have found work in the last 12 months.

If possible, try to discover an historical figure or set of figures with which to compare the ones you are plotting.

As another example, the average wage of a steel worker in West Germany is $13.45 per hour. Seems high? It is $23.90 in the United States and $2.39 in South Korea. It might be interesting to compare the output of the steelworker in those three countries. Then the real picture of productivity can be seen. Does the expansively paid U.S. worker produce more per year than the meanly paid South Korean? Perhaps not. The responsibility of the chartmaker extends beyond the simple plotting of figures. You should at least suggest that the whole picture cannot always be grasped with one set of figures.

Try not to take figures out of context. On the contrary, try to place them in a context that illuminates them even more.

How Many Charts Are Needed to Present the Information?

The short answer is: As many as are needed to avoid complication.

A chart that is too busy or too stuffed with information begs two questions. First, is all the data necessary? Second, if so, why not split it into two charts or don't use a chart at all? It is just as dangerous to oversimplify as it is to put too much in. The overcomplicated chart will not win as many readers, and those who do attempt to understand it run the risk of being confused. What you as the chartmaker should understand is whether or not by simplifying the information you have made whatever point there is clearer. Almost always this will be true, especially for a lay audience who will not take the trouble to unravel a complicated, but not necessarily incorrect chart.

There is one clear case when a series of charts is better than a single one. When making a slide presentation, the amount of information given on each slide must be kept to a minimum. It is harder for an audience listening to a lecture and viewing slides to assimilate all the information coming into their ears and eyes at once, than it is for a magazine reader, who is in control of the amount of time he or she will give to studying the information. Slides therefore should be bolder and make only one point and that as simply as possible, or the audience will not retain the information. Furthermore, even if you keep your slides simple, you would be well advised to prepare a printed summary of the salient facts for the people attending the meeting/lecture to take away at the end and study in their own time. They cannot take notes in a darkened room.

When to Avoid Using a Chart?

Do not be afraid to suggest this as a solution to the problem presented to you. You will save everyone a lot of time if after studying the raw data you can see that it is (1) too simple to even bother with, (2) so complicated that even splitting it up into more than one image will still not explain the material, or (3) you can encapsulate the essence of the information more easily in a sentence than in a graphic translation.

As the vogue for charts gains speed and acceptance of them increases, there will be more and more occasions in which a chart is asked for when it is not really appropriate.

Avoid doing a chart when it is really only fulfilling the role of decoration or of making a page or presentation look more authoritative, factual, or important. Some daily papers are using too many graphs, charts, and other graphic bits and pieces to try to attract readers. When all information is presented graphically, none of it serves the purpose that it should. It is fighting with itself for the reader's eye.

While anyone working in this field must applaud the greater use of charts nowadays and must be interested in anything that promotes the use of graphics as a means of disseminating information, there is a fine line between informing readers and annoying them with countless visual metaphors for inflation, recession, squeezes, soaring rises in prices, precipitous falls in output, sudden spurts in stocks.

It is difficult to estimate the time for each of the methods described, but in strictly production terms, and not counting the amount of time used to think about the problem and discuss the importance of this factor over that, how much time you should allow is shown here.

One-Time Methods

By hand: 10 to 20 minutes

With Letraset and other graphic aids: 2 hours

Slides: 24 hours

Overhead projection: After 20 minutes of preparation, as long as the live presentation itself takes

Computer printer: 2 to 4 minutes after feeding in the information

Graphic calculator-printer for plotting: 5 minutes, including feeding in the information

Production Methods

Flat art for black and white: 1 to 2 hours

Flat art with multiple overlays for color: 2 to 3 hours

Computer-generated color and graduated backgrounds: 3 to 4 hours (after keyline drawing is scanned in)

Complete computer-generated chart (to graphic printing quality): 7 hours

All the above depend upon the complexity of the image being drawn. A simple bar chart will of course take far less time to produce in any of the methods than a chart with an intricate drawing as part of it.

STEP 4. FIND THE RIGHT SYMBOL

If you belong to the school of people who believe that charts should only present statistics in the most straightforward, plain way, with no other visual help to the reader, for example, than the bar of the bar chart, the line of the fever graph, the circle of the pie chart, or the rules of the table, then move on to another part of the book.

As long as the artist understands that the primary function is to convey statistics and respect that duty, then you can have fun (or be serious) with the image: that is, the form in which those statistics appear. "Boredom is as much a threat in visual design as it is elsewhere in art and communication. The mind and eye demand stimulation and surprise." This statement from *A Primer of Visual Literacy* by Donis A. Donis sums up the point. While some charts that appear daily in newspapers should be very straightforward and simple—precisely because they appear every day—most of the rest can do with some visual help.

How to Look for Ideas: The Scrapbook Approach. Whenever you see an image in a magazine or newspaper or even on a packet of cereal on the breakfast table, cut it out and keep it in a scrapbook. My own method is to put whatever I find into the book in no particular order, just lightly taped so it can be removed if necessary. In this way, when searching for an idea with a blank mind, all sorts of images can be viewed that will start the mind working—and perhaps along different lines from the preconceived. Even when you know what it is you have put into the book at some time and are simply trying to recover that piece, other images will be flicking past your eyes as you search for it. The brain has a great capacity for putting odd images together that it might not have thought up without this visual stimulus.

The scrapbook, then, should be a repository both for an artist's actual reference and for interesting images that may someday be useful, though at the time of collection you may not know exactly what for. The sort of things that you should keep for actual reference are side views of ships, planes, tanks, photographs of supermarket trolleys (great for consumer prices charts), houses of all types, watch faces, elephants, donkeys, and so on—all the things that you will have to draw at one time or another. The other category—the free-form, association-of-idea category—could include anything that suggests an idea: speed, achievement, climbing, falling, reaching, hiding, all the things that statistics do. You may include photographs, drawings, paintings, advertisements, pages from catalogs, or photographs you have taken yourself.

When used for the chart, one factor will be common to any and all the symbols you choose: they must be instantly recognizable. They must be familiar.

Sports. If you are doing a chart about the performance of a company consider a sports analogy: jumping over, running past, getting to the line first, hurdling, diving into, galloping by, scoring a goal, winning the serve. Sports competition is instantly recognizable as a symbol of achievement or winning.

Tools. Tool catalogs are full of wonderful images that can crack down, bring pressure on, cut, saw, or chop in half, squeeze tightly, measure exactly. The sight of the tool immediately conjurs up the action that it is used for, and that association with the statistics can then follow naturally.

Domestic Things. Vacuum cleaners that suck away profits, refrigerators that freeze prices, trash cans that contain last year's budget, scales that measure how this year's budget is balanced, beds for a sick economy, plants for a growing economy, coffins for a dead one, windows for a look through to the future, books for reading of the whatever-it-is industry are all everyday household images that can be drawn upon.

Animals. Symbols that are always associated with events should at least be considered. The bull market (for an upward surge) and the bear market (the other way) are standard symbols, as are hawks and doves for things militaristic and peaceful, the Republican elephant and the Democratic donkey, the eagle for all things American and the bear for the Soviets.

The deliberately relaxed and random quality of a scrapbook containing anything you find interesting will trigger all sorts of visual associations and help in the creation of images.

(Overleaf) A collection of cuttings from a scrapbook including photographs, symbols, your own (or your children's) sketches, stamps, catalogs, old engravings, and parts of advertisements can be used either as reference or as they are.

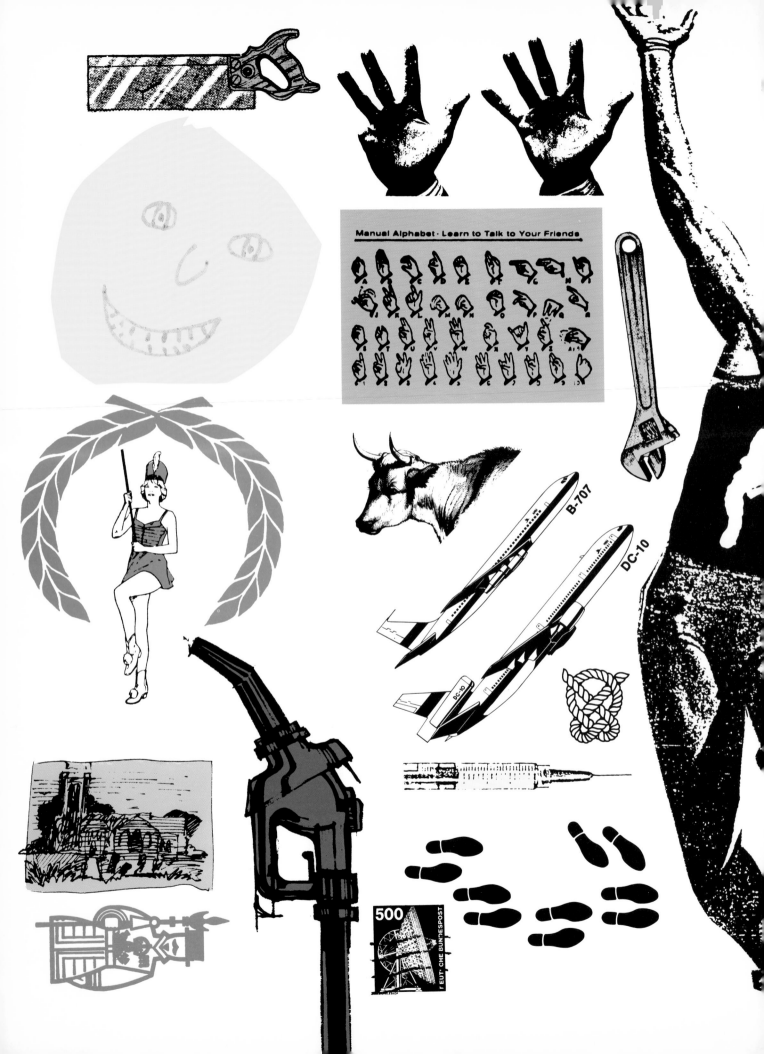

Manual Alphabet · Learn to Talk to Your Friends

B-707

DC-10

DC-10

500

EUT CHE BUNDESPOST

Economic bulls and bears and political elephants and donkeys can be found in old printers' catalogs, or can be specially drawn to make a related pairing, as in the case of the Republican and Democratic symbols. The idea of the animals may also suggest a form for a chart: thus donkeys kicking back, elephants lumbering along heavily, bulls snorting or charging, bears snarling, clawing, or hugging at the figures.

Having fun with the figures: In this fever chart about the U.S. birthrate (the yearly number of births per 1,000 women), the exaggeration of the size of the diaper provides a good background for the information and also looks humorous. Note that the fever line is the heaviest part of the drawing and that the baby is sketched with the simplest of thin lines.

And the attributes that animals carry with them are also widely known. Slow growth is synonymous with the tortoise; fast growth with the racehorse, hare, or cheetah; a prickly problem with the porcupine. Some animals are walking charts—the zebra with its bars, for instance (although it is hard to think of anything that it could be applied to except statistics about zebras). Put it into the scrapbook anyway. Who knows when it might be just right?

Fun. Humor is a great weapon in your visual arsenal. As long as it is not malicious, making people laugh with you will usually help them remember your image and therefore the point of the chart. Even a smile will encourage a reader to look into the statistics he or she might not have thought of reading in a less embellished chart.

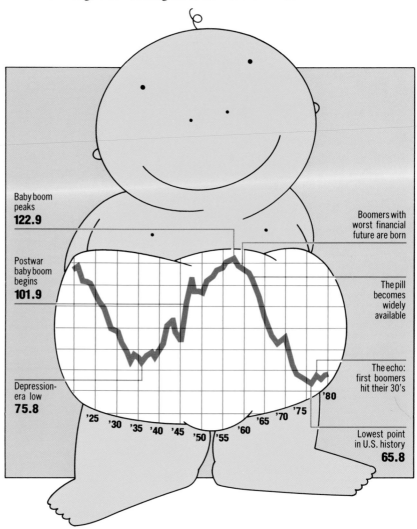

Figures of Speech. "The tail wagging the dog," "The almighty dollar," "Painting the town red," "Down the drain," "Over the hill"— familiar turns of phrase can be literally drawn. This will assume a certain literary sophistication on the part of the reader, but it can be most effective. The assumption is that the reader knows the illustrated phrase. You might help your audience along by using it in the chart's title.

However, it must be a phrase that once grasped has some meaning within the context of the chart. Do not use an illustrated phrase just for effect. Whatever the phrase implies must back up the figures, not be purely arbitrary. In the accompanying illustration it did seem as though the oil-producing countries had overpriced their product, produced too

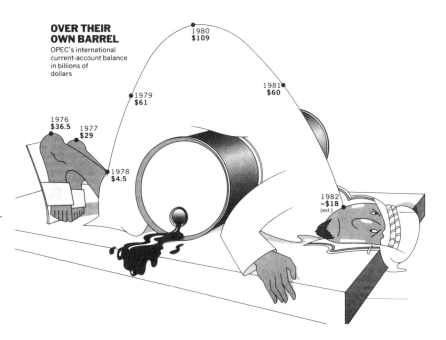

OVER THEIR OWN BARREL
OPEC's international current-account balance in billions of dollars

1980 $109

1979 $61

1981 $60

1976 $36.5

1977 $29

1978 $4.5

1982 −$18 (est.)

Figures of speech can be literally illustrated. The phrase itself usually makes the best title for the chart, especially if it is changed slightly to fit exactly the sense of the idea, in this case "over a barrel."

much of it, and thus were in a very difficult position. In fact being "over a barrel," was, in this case, their own dilemma.

What Criteria Apply When Screening Possible Images?

The first, as discussed above, is a clear understanding of the symbol. There must be no doubt at all about the meaning behind the use of the symbol in the context it is used. If the reader misses the point, it would have been better to have used no symbol at all. Ask your friends or colleagues if they understand the chart you have drawn. Inquire if they understand something else from the symbol you are showing them.

Do not be surprised if a colleague from abroad gives you an answer you hadn't expected. While symbols and the study of their meanings fill whole books, a few attempts have been made to create a universally understood language of symbols—a system that people of any country could understand—but few are instantly recognized around the whole world. Make sure, for instance, that the readers of your chart are not from the Far East if you want to show the underneath of a foot for some reason. It is considered extremely bad taste to do so in certain parts of Asia. Be aware that Australians hate being characterized as kangaroos and that you will never work again in France if you draw them as frogs.

The question of taste is largely personal, and there will usually be a system of censure built into the client's checking process so that the alarm can be raised before you spend long hours lovingly drawing.

But you must exercise your own common sense and sense of decency and taste in these matters, and you should have a fairly well-balanced view of what is going to be offensive to any of the prospective readers of your work. Particularly sensitive areas are the depictions of different nationalities, where the cartoonist's time-honored stereotypes may not be appropriate. If you need to differentiate between countries it may be better to use a flag rather than attempt to draw a person from that country. The chartmaker is seldom in the same position as the editorial cartoonist, who has the freedom and license to be as sexist, racist, and biased as he or she wants within his or her own bounds of taste. Those bounds probably will be set a good deal further away from the straight and narrow than the chartmaker's ever can be.

Avoiding Symbols. As is the case when a chart should be totally avoided, there will also be times when no symbol is appropriate, or to

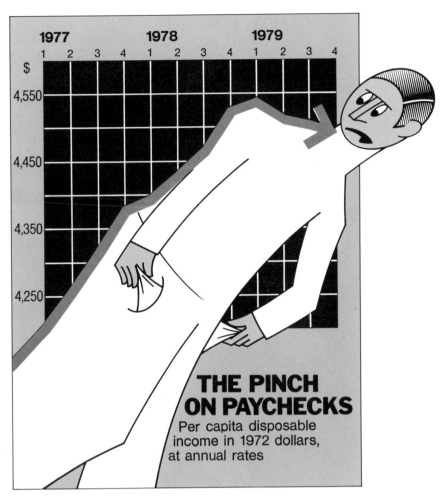

**THE PINCH
ON PAYCHECKS**
Per capita disposable
income in 1972 dollars,
at annual rates

A decline in the amount of disposable income shown like a snake attacking the throat of some unfortunate individual allows the reader to smile at a situation without losing sight of the facts. The idea of what is happening is conveyed at the same time as the information. Humor will help it be remembered.

put it another way, any symbol is inappropriate. The extremely distressing nature of some illnesses, the rise or fall of which you may be charting, should be plainly left to a simple graphic, abstract image. Fashions in taste change, but while the statistics may be most necessary, an illustration or visual metaphor to go with the statistics should sometimes be left out.

The final criteria goes back to the first step in this chapter. Apply the correct level of symbol or illustration to the audience. Do not visually speak down to sophisticates or over the heads of children.

What and Where Are the Best Sources?
Dover Pictorial Archives has an excellent catalog of their publications. These contain out-of-copyright illustrations that can be used by themselves or as a starting point for an illustration or an idea. The range includes borders, lettering, signs and symbols, historical ornament, folk art motifs, animals, music, food and drink, and early advertising art, to mention only a few from a list of about 150.

Perhaps the most immediately useful source is *The Handbook of Pictorial Symbols* by Rudolf Modley, which has masses of black and white simplified drawings displayed very clearly. The only word of warning: Some of the pictures are a little dated now, but they can easily be updated with a few judicious changes here and there. Some fairly difficult concepts are included, such as the five graphic ways of committing suicide. It also has a section of public symbol systems, such as those you see at airports, exhibitions, and the Olympic games. It is very instructive to see how a number of other designers have solved the same problems: even what might be considered such a simple one as the depiction of a telephone.

Pictorial symbols can be found in a number of different places from catalogs to specific collections in books. Make sure that they are copyright free before using them directly from your source.

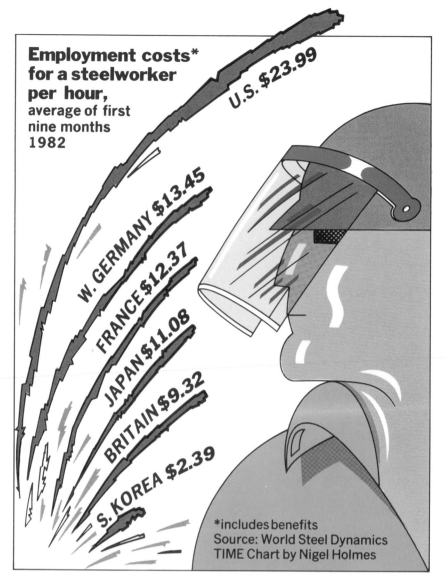

**Employment costs*
for a steelworker
per hour,**
average of first
nine months
1982

U.S. $23.99

W. GERMANY $13.45

FRANCE $12.37

JAPAN $11.08

BRITAIN $9.32

S. KOREA $2.39

*includes benefits
Source: World Steel Dynamics
TIME Chart by Nigel Holmes

Illustrations from encyclopedias, annual reports, magazines, or other printed material are great sources. However, they should be merely a starting point for your own image. Too close a representation of a photograph can result in awkward legal proceedings!

Other sources are the catalogs mentioned in the Selected Bibliography and the idea that anything you see or hear might someday be useful, whether you see it on TV or flashing past you on a train or hear it in a conversation with a friend.

Letraset and the other graphic aids companies have catalogs of their own symbols, which can be useful in themselves. Without actually buying the sheets of press type, you can use them as a starting point from which to draw. If you want to use their products, it is a wasteful process to try to photocopy them from the catalog to save the cost of buying the sheet; the quality is never the same.

Do not be afraid of using the obvious. If a chart is about money, use money symbols. If the chart is about a number of people, use pictures of people. If the chart is about the number of cars produced by the car industry, use cars.

Look up in a pictorial encyclopedia anything that you can think of that is even marginally related to your subject. For instance, if you are doing a chart about steel production see what the encyclopedia has. There will probably be an illustration of how steel is made, what the finished bars look like, what a steel worker looks like, and how the furnace works. Some of these may provide background ideas or even entire images. For instance, the accompanying chart about a steel worker was inspired by an image in an encyclopedia, with the sparks flying toward the worker, here representing the cost of his employment.

Chapter Four
Nine Assignments

THIS CHAPTER PRESENTS nine assignments from the initial problem and receipt of data to the finished graph, chart, or table. Each example will be taken from a statement and discussion of the problem and through the four steps described in the previous chapter: identify the reader/user, select a production method, review the numbers, and find the right symbol.

The assignments were selected to represent the diversity of jobs that a chartmaker might encounter. To give each one a specific focus, not only the job itself, but also the forum in which the final product will appear has been decided upon before the start of each project. This way different types of chart can be shown in different but typical situations.

The accompanying table lists in the first column the type of information to be presented (time sequence, budget, work progress, and so on). Across the top are the areas or forums where the finished piece will be used (for instance, in a daily paper or magazine, as a teaching aid).

The numbers refer to the nine assignments. Where the same number appears twice or more on the chart, alternate versions of the assignment have been prepared to show adaptions or completely different forms derived from the same data.

	Daily Paper	Magazine	Business	Information Aid	Teaching Aid	Personal Presentation
Time Sequence	**1** Fever	**1** Fever	**2** Bar			
Budget		**3** Pie	**3** Pie			**3** Pie
Overlapping Time Schedule				**4** Bar		**5** Bar
Accumulation	**6** Bar	**6** Table				
Organization			**7** Table		**7** Table	
Matrix				**8** Table	**9** Table	

OPEC's Balance of Payments 1976-1982

PROBLEM

Show how OPEC's current account has behaved during recent years, ending with a massive loss estimated for 1982.

DATA

These are given in billions of dollars.

1976	$ 36.5
1977	$ 29
1978	$ 4.5
1979	$ 62
1980	$115
1981	$ 67
1982 est.	−$ 18

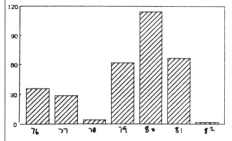

A quick printout from a small computer/calculator helps you see the shape of the numbers.

INTENDED USE

This chart will be prepared to appear in (1) a daily newspaper in black and white and (2) the news section of a magazine in 4 color.

DISCUSSION OF THE PROBLEM

In the first months of 1983, the cartel of oil-producing countries was arguing among themselves over the amount of oil they should be producing and what they should be charging for it. Since the price of oil had increased, buyers had found ways of conserving energy, being able to do with less or, in the case of America, taking steps to become more self-reliant by increasing domestic production. There was a feeling that OPEC had overstepped the mark with their prices and now in a world market where demand and consumption were declining that OPEC had brought their troubles upon themselves by being too greedy. They were

Newspaper stories fill in the necessary background information on this assignment.

flooding the market by producing too much oil, but not enough people were buying it since it was too highly priced.

Some members of the organization argued that the price should be left as it was and production should be cut; others argued that the price should be cut while production kept at the same level. At one point in the meetings it seemed that the cartel might be about to break up. At the very least it was clear that the countries had lost much of their power. America certainly had found that it could do with less, and now if they who had caused people such stress and gas lines were themselves in trouble, well, they deserved it.

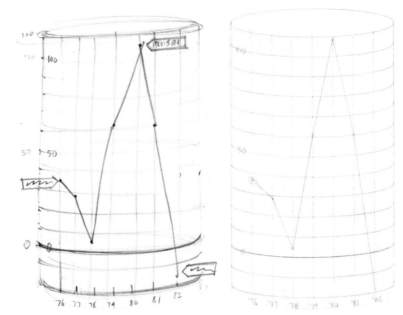

(Left) The rough idea is sketched out; then the figures are accurately plotted on tracing paper (right).

STEP ONE

Identifying the Reader/User

Since this is appearing in the forum of a daily paper or news magazine it can be safely assumed that it will *not* be an isolated piece of graphic design without the backup of articles on the subject. The user of the chart therefore will be able to read something about the context of what is being discussed here. If the person is a regular reader he or she will also have an accumulated experience about the information and will probably also have been affected in some way by the action of OPEC in raising its prices. Thus, most readers will have a sense of the situation before seeing the chart, be knowledgeable about the subject, and more than likely have some strong feelings about it. While the audience for the chart in the newspaper may well be exactly the same as that for the magazine, the difference for the chartmaker is how that same audience reads information in a newspaper compared with that in a magazine.

The basic difference is the speed of reading. A chart in a newspaper, especially a daily, should be direct and to the point, with less visual play than can be used in a magazine. By its very nature a magazine is generally read more carefully, and certainly it stays in the home or office longer. Thus it can take a more interpretive approach to statistics.

STEP TWO

Select a Production Method

In this case two production methods will be used: for newspaper as well as for magazine use.

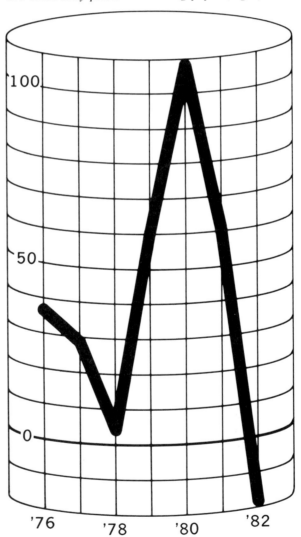

The simple final line version for use in a daily paper relies on a minimal oil barrel motif to lift the chart out of flat two-dimensional plotting.

1. Newspaper Use: Since it is stipulated that this version will be in black and white, there is little decision making here: A piece of flat art will be produced, perhaps with separate overlays for a tone or two.

2. Magazine Use: Here flat art with multiple overlays representing the colors is the correct way to produce the artwork.

In both cases before proceeding consider using a plotting calculator to give a fast look at the "shape" of the figures.

STEP THREE

Reviewing the Numbers

Although the data is available from 1970 onward, the figures from 1976 give a clear picture of the rise and fall of OPEC's fortunes, and by choosing fewer numbers there will be more room to display them. So the first thing to do is to suggest that the chart start in 1976.

The next question to examine is whether or not this set of numbers alone is enough to tell the "story" behind the information they contain. Some discussion should take place about the possible inclusion, perhaps in another chart, of figures that show OPEC's total production of oil and the world demand for it.

As far as the first is concerned, it is interesting to note that OPEC no longer has the lead in the production of oil in the world. In 1982, for the first time, non-OPEC nations produced more than the once mighty cartel. Armed with this information, an editor

might choose to add a sentence about it in the story or make more space available for a whole chart on the subject. It could even be connected to the revenue numbers by using a different scale on each side of the chart, one for amounts of money and one for barrels of oil.

As for the world's consumption of oil, a case could also be made here for a separate chart showing how the demand has dropped in the last four years.

Although neither of these last two sets of figures were presented as the information to be included in the chart at the outset of this assignment, it is important that the artist doing this job should at least be aware of the context of his or her own work, in terms of both the political background and the statistical surroundings. By taking into account other facts that relate to the job at hand, he or she will be able to make a real contribution to the reader's understanding of that work.

In this case let us assume, however, that the newspaper is tight for space on that day and gives the artist a space that is 2 inches wide by 4 inches high—a fairly standard one-column chart. The magazine has more space, but a different shape: 6 inches wide by 4 inches high.

STEP FOUR

Finding the Right Symbol

At the start of any job that has a clear visual vocabulary, it is worthwhile to make a "shopping list" of available symbols related to the subject. Thus in the case of oil, before any considera-

tion about what the figures mean, or what the chart is trying to convey, the accompanying symbols spring to mind.

They can be drawn specially for this occasion or can be found in source books or in the artist's personal scrapbook of likely images.

SOLUTION TO THE PROBLEM

In this case the problem prompted the following solutions:

1. Daily paper: A straightforward, uncluttered image will always give the best reproduction on newsprint. Plot the numbers out, at the same time thinking of which symbols from the shopping list would most speedily suggest the idea of oil to a reader. If you apply your plotted numbers to the side of an oil barrel, the message is immediate and obvious. Keep the barrel simple. There is no need to be fussy with details. A barrel is a cylinder. When the cylinder has the word "oil" on it, it becomes an oil barrel.

2. Magazine: If you are prepared to defend your ideas in the face of what may well be an unamused editor, then this set of statistics could provide you with the occasion. The fact that a once-powerful organization is in trouble can give rise to a number of images, which may first start out in the mind as words or phrases: in hot water, in quicksand, on shaky ground, sinking, impotent, and so on. Do some quick sketches to go with these ideas. What about that poor man in a barrel? How about over a barrel—their own barrel. . . .

First drawings of the idea for magazine use are needed to make sure it will work. They are rough and should be done very quickly. The sketch in color can be used to show an editor the idea and approximate coloration.

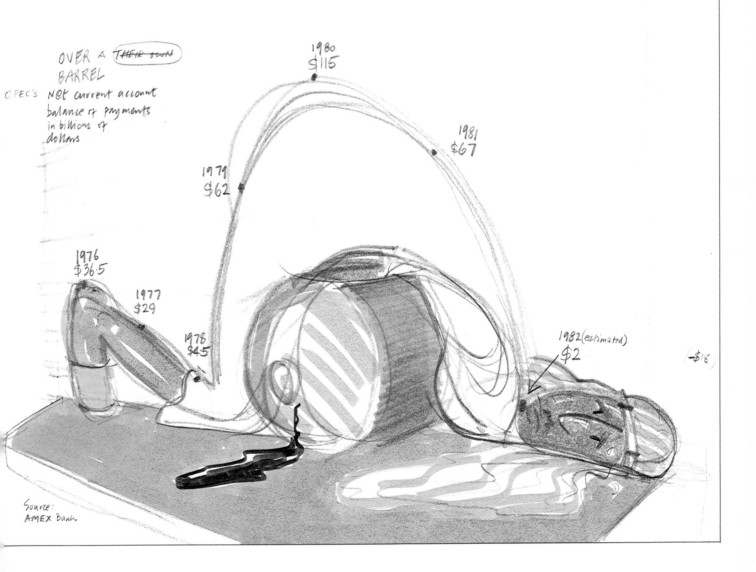

OVER A ~~THEIR OWN~~
BARREL

OPEC's Net current account balance of payments in billions of dollars

1976
$36.5

1977
$29

1978
$45

1979
$62

1980
$115

1981
$67

1982 (estimated)
$2

-$18

Source: AMEX Bank

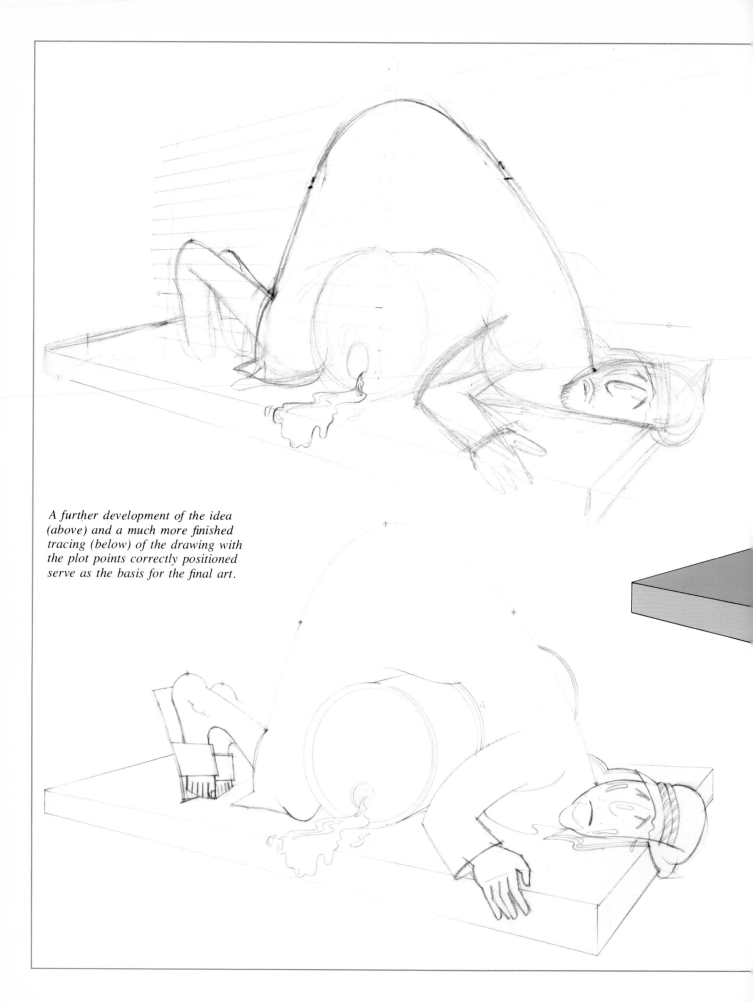

A further development of the idea (above) and a much more finished tracing (below) of the drawing with the plot points correctly positioned serve as the basis for the final art.

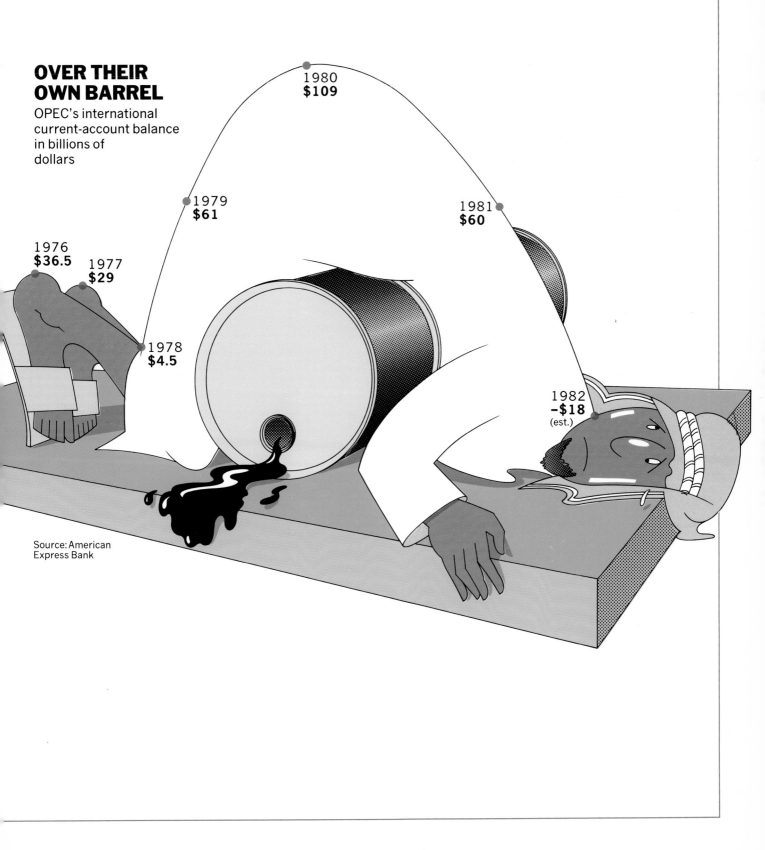

OVER THEIR OWN BARREL

OPEC's international current-account balance in billions of dollars

1980
$109

1979
$61

1981
$60

1976
$36.5

1977
$29

1978
$4.5

1982
−$18
(est.)

Source: American Express Bank

87

The Dow Jones Index of 30 Leading Industrial Stocks

PROBLEM

Show the high points, low points, and closing points of the Dow Jones Index for each day of the month in April 1983.

DATA

The following table shows all the data that should be charted.

1983			
April	*High*	*Low*	*Close*
4	1133.03	1115.13	1127.61
5	1137.58	1116.39	1120.16
6	1122.48	1102.17	1113.49
7	1122.36	1106.71	1117.65
8	1128.87	1109.52	1124.71
11	1144.54	1127.23	1141.83
12	1151.80	1134.09	1145.32
13	1165.34	1142.22	1156.64
14	1171.05	1146.67	1165.25
15	1177.05	1159.06	1171.34
18	1187.89	1164.57	1183.24
19	1187.98	1166.22	1174.54
20	1197.56	1170.47	1191.47
21	1202.11	1180.92	1188.27
22	1204.62	1184.21	1196.30
25	1206.75	1183.24	1187.21
26	1210.82	1177.92	1209.46
27	1225.04	1197.37	1208.40
28	1226.97	1202.04	1219.52
29	1235.49	1208.59	1226.20

INTENDED USE

The business section of a magazine is where the chart will appear.

DISCUSSION OF THE PROBLEM

After the market rose almost continually for eight months, experts were wondering at the beginning of April 1983 whether the market might be ready for a sharp drop. But with the added impetus of large amounts of IRA investment money, the bull market continued to roar ahead. Having broken the 1000 point mark in October 1982 and 1100 in February 1983, it was heading for 1200 as the middle of April

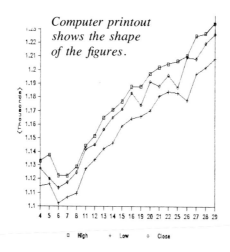

Computer printout shows the shape of the figures.

approached. It closed higher than that on April 26 and continued up.

STEP ONE

Identifying the Reader/User

If the chart is to appear in a business magazine or in the business section of a general magazine you can safely assume a reader who knows the subject. Since this chart is to occupy a small space only, it seems at first glance as though this would require a straightforward presentation of the facts. If the chart is one of a series, perhaps representing some information from the issue of the week before (in a weekly publication), you may decide to simply update last week's chart, without changing the design. This provides continuity for a reader who simply requires the information as quickly as possible, without having to encounter a separate visual idea. It also suggests, subtly, that you are taking the figures and the reader seriously by avoiding an easy chance to underline the bull market with a striking visual interpretation.

STEP TWO

Selecting a Production Method

Running in black and white, this chart will be prepared as a single piece of flat art, camera ready for the publication's normal production procedure. It is a good chance to use a computer to do the plotting.

STEP THREE

Review the Numbers

The closing figures start at 1127, dip down to 1113, and then climb up to 1226 at the end of the period in question.

To achieve some separation in the plot points, the scale should be broken, starting not at 0 but at 1100, which is the nearest round number to the lowest plot point (1102.17: the "low" on April 6). The scale must go up to 1240 to accommodate the highest plot point (1235.49 on April 29).

The value of this chart to its readers is the completeness of the information it gives. Therefore a consideration of the numbers is not an exercise here to see which figures can be omitted in the cause of clarity or design. This is not, in other words, a chart that gives a generalized trend in the Dow Jones, it *is* the Dow Jones. And it includes three numbers for each day. The most often reported number is the closing—that is, the official index number at the end of the day's trading. But of great interest to those who play the market are the high and low points that are reached during the day. They can provide clues to future market performance, and in fact analysts use charts of this type extensively to forecast the next moves on Wall Street and whether their clients should buy or sell.

The design consideration in all this is to make sure that the three points are clearly visible for each day.

STEP FOUR

Find the Right Symbol

This chart is deliberately simple visually and to the point statistically. The question is how to draw the lines so that they are clear. An unillustrated chart does not necessarily take less time to design. Because the eye is not going to be entertained in other ways, the designer must pay more attention

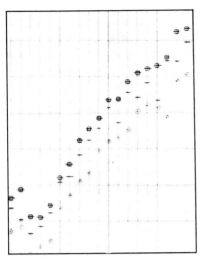

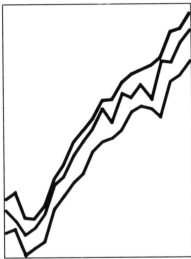

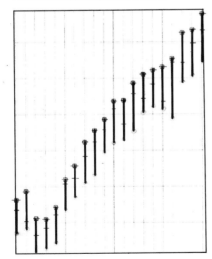

Hand-plotted working tracings show the different ways that the figures can be plotted: (top) the plot points by themselves; (middle) as three fever lines; (bottom) as double-ended bars with three reading points on each bar. The double-ended bars are preferable when high, close, and low points are all to be shown.

to the details of the weight of line, grid, numbering scale, labels, and title. Once this is done it would be worth considering the introduction of a small symbolic bull that could be placed in a convenient part of the chart where it does not interfere with the information. At the same time a bear could be designed (or selected from a book of stock images) to reflect the market changes when they occur. This little touch adds some visual relief and imparts an instant message about the state of the market without dominating the look of the chart.

SOLUTION TO THE PROBLEM

Plot out the figures, making a different mark or colored dot for the high, close, and low points. This plotting is then the basis for two different ways of presenting the information.

The first example connects the dots to form three separate fever lines, showing all the high points connected, all the closings connected, and all the lows connected.

The second example, much more commonly used, connects the high, close, and low points for one day by means of a vertical bar, before moving on to do the same thing for the next day's statistics.

While the fever lines in the first example do what any fever chart is excellent at doing, namely, indicating the broad sweep of the numbers across the time period, it is the second example that will best display these particular numbers. Since it also is the way that most people have seen stock market charts before and since the purpose here is to be straightforward and businesslike, it seems to be important to ring the bell of continuity in the reader's mind.

The vertical grid should be restricted to one line for the beginning of each week, leaving out the other day's lines, and the horizontal grid lines should be kept down to one for every 10 or 20 points on the index, with a tick mark for every single point. This scale should be set on both the left-hand and the right-hand edges of the chart. This way the maximum amount of information can be gleaned from the plotted points.

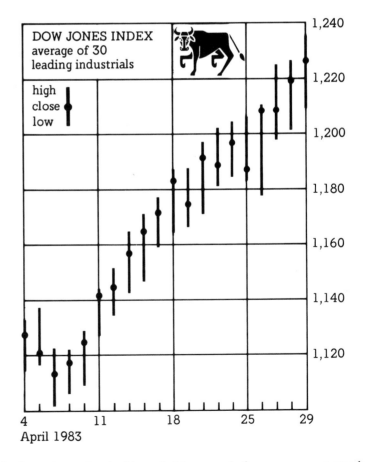

The final art shows the double-ended bars ready for newspaper reproduction.

The Budget Dollar

PROBLEM

Show how each dollar of the federal U.S. budget is broken down into its component parts. Furthermore, show where the money comes from and how it will be spent.

DATA

The figures in the following table are for fiscal year 1980 and represent the cents in a dollar.

Income:		
	Personal income tax	43
	Social Security & related taxes	30
	Corporate income taxes	13
	Excise taxes	4
	Borrowing	5
	Other	5
Spending:	Benefit payments to individuals*	39
	Defense	24
	Grants to states & localities†	16
	Other federal expenses	12
	Net interest	9

*Social Security, federal employee benefits, Medicare.

†Medicaid, education, public assistance.

INTENDED USE

The chart will be used in three different ways: (1) in a general magazine in two colors, (2) in a business magazine in black and white, and (3) as a personal presentation.

DISCUSSION OF THE PROBLEM

This is not a statistical presentation that will deal with the actual amounts of dollars in the federal budget; rather it is a representation of how the money is proportionally divided up. It could be expressed as percentages of the total amount; thus personal income tax represents 43 percent of the total budget receipts. But since those totals are measured in billions of dollars, it is much easier to understand the concept of divisions of a single dollar. In a way, it brings the figures down to earth; it gives people a chance to relate to them more easily than to the huge figures involved in the actual dollar totals.

In this case it seems perfectly natural that a pie chart should be used. To try to force some other form onto these statistics would be perverse, and it would also lose the recognition factor that an audience relies on when it first sees a statistical presentation: the ability to understand what is expected of a chart before any details of the statistics are studied. The more sophisticated an audience, the more it will already know the different types of charts.

STEP ONE

Identify the Reader/User

Each of the three different uses means that there will be different readers whose varied needs and backgrounds must be appropriately met.

1. General Magazine: A general audience would be expected to understand the concept of breaking down a budget. They may well be familiar with similar graphic representations, for instance, of how the telephone or electric company uses their money.

2. Business Magazine: When reading a business magazine the audience will generally expect to be treated differently, although they may be the same readers, in some cases, as the general magazine's. In this case they are reading more as specialists than as an interested general public. Therefore, the same data should be presented in a somewhat more straightforward, businesslike way, so as not to appear to be patronizing. There is criticism of overdesigned or overillustrated and embellished graphics, and it is well founded criticism if the charts in question are published in specialist or technical journals. Where an audience knows its business, do not dress up the facts.

3. Personal Presentation: Let's assume that a budget director needs to present a set of figures like these to the President and chief executives of the company. This of course is the most specific of all audiences—only a few people who presumably know all about their company in the greatest detail. (The purpose of such a presentation, for instance, may be a preliminary stage in the preparation of their annual report.)

STEP TWO

Select a Production Method

Three different methods must be chosen to accommodate the different uses.

1. The general magazine will need flat art with overlays. Black and one other color are to be used, which means several overlays will be necessary.

2. The business magazine requires that the art be produced the same way. But because it has decided to run the chart in black and white, less overlays will be needed.

3. The personal presentation is to be held in the company's conference room. Only a few people will see it. It could therefore be a slide or an overhead projection display, or simply a large card with the chart drawn on it. These are all one-time production methods. Since there is only one set of statistics to show, it seems unreasonable to have to set up a projector and screen and dim the lights all for one slide. A better idea is to make a large one-time chart that can be held up at the meeting for all to see. Reduced photocopies of it can be made if time permits.

STEP THREE

Review the Numbers

Since a pie chart seems appropriate for this problem, a review of the numbers confirms that this will be possible. No numbers are so small that they will appear as insignificant or annoyingly thin slivers of the pie. There are six categories in the income part and five in the spending part: both ideal numbers of division for a clear presentation in a pie chart.

A question should always be asked

about "other" categories, especially if that division of the whole is larger than any of the rest. Usually the "other" categories contain so many smaller amounts that it is impossible to show them all without confusion. However, what are the "other federal expenses" on the spending side that account for 12 percent of the total? When one realizes that the total spending was $577 billion for that year, a slice of the pie representing $69 billion and labeled "other federal expenses" needs at least to be looked at!

STEP FOUR

Find the Right Symbol

As with the assignment about oil, where visual symbols come to mind independently of any consideration about the nature of the figures involved, a chart about budgets also suggests symbols. "Budget" and "balancing" seem to go together, and the fact that this chart is about government spending increases the range of symbols worth considering.

1. General Magazine: As described earlier, there is more latitude here for illustrating the figures, provided enough space has been allotted to the chart. A combination of the see-saw with income at one end of it and outgoing payments at the other shows the balances between the two elements and could even be tipped in one direction to indicate that more was being spent than received (thus producing a deficit).

2. Business Magazine: The more conventional approach here and smaller space available restrict the designer in choice of symbol. Perhaps the capitol building with two pie charts superimposed at either end is all that is necessary to show the facts and give a little informative visual relief.

3. Personal Presentation: In this case the essence is speed and fast communication of the facts. Because other breakdowns of figures are continually being presented to this audience, they need to know instantly that the chart they are being shown is about money, not, for example, job distributions or amounts of their product carved up into geographical areas. So a money symbol of some sort should accompany this hand-drawn chart. You might even consider using a real dollar bill to illustrate the point.

SOLUTION TO THE PROBLEM

Draw out the two pie charts to give you an idea of how the type will, or will not, fit into various segments of the circle. If you are working larger than the finished printed size, remember to consider the size of type when printed.

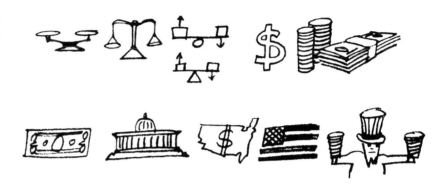

1. General Magazine: Place the two pie charts on the balanced bar. At this point consider whether or not to include the subtle idea of overbalancing the see-saw on the spending side, thus indicating a budget deficit.

In order to link this visually to the government, turn the two pies into

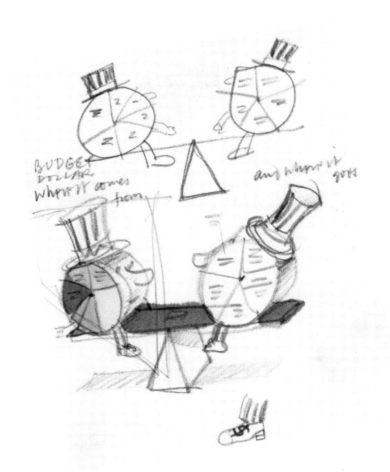

Rough sketches, however small and quickly drawn, are useful ways of making sure that the idea works. They also show you the kind of spaces that you are going to have to fit the type into.

one overall title, "The Budget Dollar," is immediately qualified by the two subtitles, which are arranged to be read as one sentence while still applying to the two separate pies. It is important also to make room for any simplified Uncle Sams by adding another dimension to the circles and then drawing in legs, faces, and most obviously of all, the instantly recognizable top hats with stars and stripes.

Organize your picture within the space to allow clear titling. In this case footnotes needed to amplify the labeling of the segments or slices of the pie.

2. Business Magazine: Having already drawn out the pies, it is now simply a case of arranging them together with the capitol building symbol and all the necessary titles, subtitles, and notes within the space available.

There is something to be said for placing the capitol building at the bottom of the chart on the "ground" as it were. This has the effect of stabilizing the design, anchoring it to something that readers intuitively recognize as correct. Of course to carry the analogy through means that pie charts are floating in the "sky," ominously about to collide like two huge planets in space, but that might not be too bad a subliminal thought to pass across to the reader. In any case, it seems better to have the building on the ground where it naturally would be and the two abstract representations filling the rest of the space, whether that space is seen to be "sky" or not.

As a secondary consideration, the ground "landscape" is a neat enclosing device for the subtitles to the two pies themselves, and by turning them into arrows, the sense of the word inside them is visually explained. Thus income is shown to be coming into the

building and spending is leaving the building. Again, enough space is left for the important qualifying notes.

3. Personal Presentation: However small the pies are when drawn, they can very simply be enlarged by projecting the lines out to as large a circle as is necessary for a presentation. Even if the pie charts are drawn by a computer (see pages 63–64), this method can be used to enlarge them. If a computer or graphic printer has been used for plotting, see what its output is for a single bar divided into percentages. This might be a more convenient way to show the figures when drawing the chart by hand. Certainly, it allows as much room as you need to put down the words, since they all have to be alongside the bars. Draw it at a scale at which the dollar bills fit neatly over the two columns; then place the real bills in position.

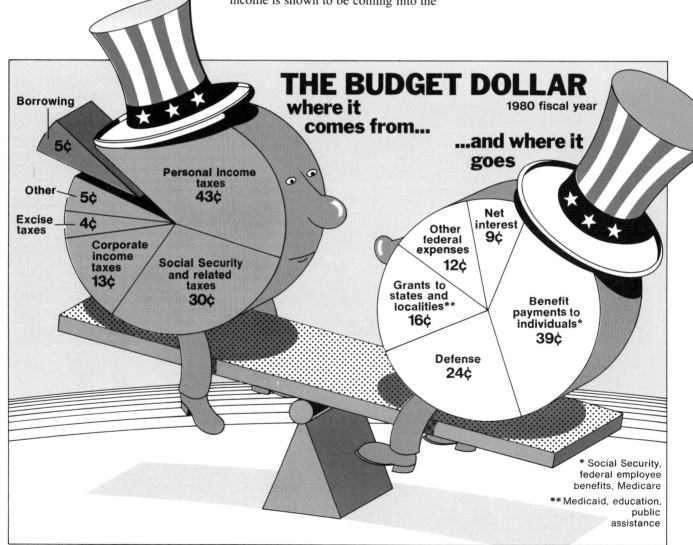

THE BUDGET DOLLAR
where it comes from...
1980 fiscal year

...and where it goes

Borrowing
5¢

Other 5¢

Excise taxes 4¢

Personal income taxes 43¢

Corporate income taxes 13¢

Social Security and related taxes 30¢

Other federal expenses 12¢

Net interest 9¢

Grants to states and localities** 16¢

Benefit payments to individuals* 39¢

Defense 24¢

* Social Security, federal employee benefits, Medicare

** Medicaid, education, public assistance

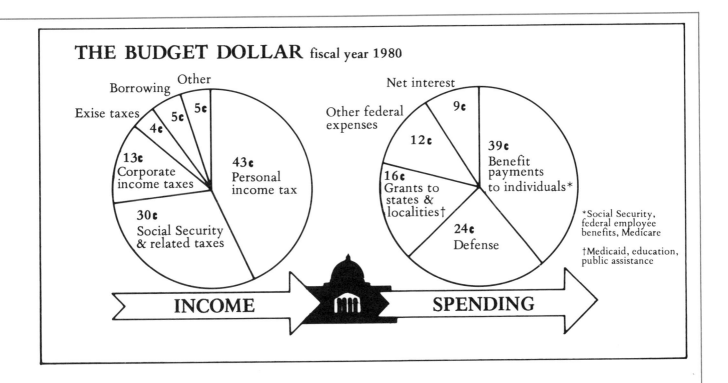

THE BUDGET DOLLAR fiscal year 1980

INCOME

- Other 5¢
- Borrowing 5¢
- Exise taxes 4¢
- 13¢ Corporate income taxes
- 43¢ Personal income tax
- 30¢ Social Security & related taxes

SPENDING

- Net interest 9¢
- Other federal expenses 12¢
- 39¢ Benefit payments to individuals*
- 16¢ Grants to states & localities†
- 24¢ Defense

*Social Security, federal employee benefits, Medicare

†Medicaid, education, public assistance

(Left) In the finished 2-color magazine piece note the slice in the left-
hand "Sam" that represents borrowing. The insertion of this part will set
the balance right.

(Above) The final black-and-white line
version is for newspaper use.

(Right) For the personal presentation, the "vertical pie chart," or bar
divided into percentages, is used including real dollar bills. The words
were enlarged from type set on a
typewriter with interchangeable
faces.

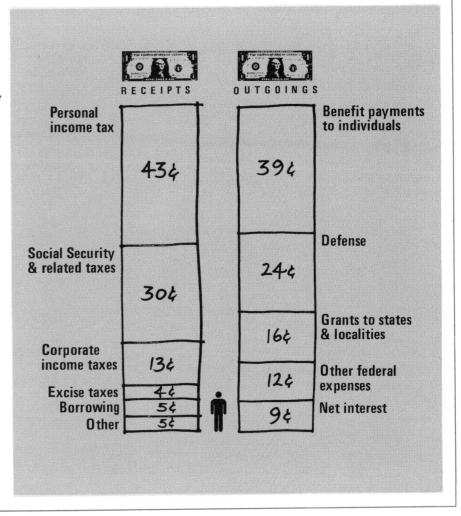

RECEIPTS

- Personal income tax 43¢
- Social Security & related taxes 30¢
- Corporate income taxes 13¢
- Excise taxes 4¢
- Borrowing 5¢
- Other 5¢

OUTGOINGS

- Benefit payments to individuals 39¢
- Defense 24¢
- Grants to states & localities 16¢
- Other federal expenses 12¢
- Net interest 9¢

Orchestra Guide for Concert Goers

PROBLEM

Show the historical development of modern orchestral instruments from 1650 to the present day, as well as the lives of the great composers beginning at the same time.

It can be very revealing to track the history of two different types of information alongside one another—for example, on the one hand methods of transport, with methods of communication on the other. Other examples are reigning English monarchs on one side, with the wars being waged by England on the other; inflation rates and U.S. Presidents; jazz movements and art movements. At the very least, the one provides a context for the other: What was the U.S. inflation rate during Kennedy's presidency? What was Picasso painting while Louis Armstrong was recording "Tiger Rag?"

The progress of technology has enabled artists to develop their art. In some cases it is even the starting point for their art, and the chart in this assignment is an attempt to show concert goers a relationship between the invention of musical instruments and their use by composers of the time.

DATA

The following data on composers' lives and the development of orchestral instruments will be compared.

INTENDED USE

This will be used as an insert into the programs for a series of orchestral concerts, not related to any one in particular, but rather as a general piece of information.

DISCUSSION OF THE PROBLEM

There is never any harm in knowing more about a subject, even in the relaxed surroundings of a concert hall or an art gallery or a jazz club. This chart gives additional detailed information about the evening's music; it is not an essential part of the concert, but a welcome bonus. The readers are left to draw their own conclusions about the sort of music that any of the composers might have written had they lived at different times, with correspondingly more or less instruments in their musical palette. In that sense this chart offers no point of view—merely the information.

STEP ONE

Identify the Reader/User

A largely adult audience attends the concerts at which this chart will be seen. They are also interested in the subject of the chart, since it is directly involved with their evening's entertainment. They are also a captive audience, at least for the duration of the concert. For those few who are not so well acquainted with the music, the composers, or the instruments, this chart will help them enjoy the program by understanding a little more about the history behind it.

STEP TWO

Select a Production Method

Since this will be printed in full color, the art will be produced by drawing the basic lines on drafting film and using overlays for the colors.

STEP THREE

Review the Data

The comprehensiveness of the infor-

Composers' Lives

Purcell 1659–1695
J.S. Bach 1685–1750
Handel 1685–1759
Haydn 1732–1809
Mozart 1756–1791
Beethoven 1770–1827
Berlioz 1803–1869
Chopin 1810–1849
Schumann 1810–1856
Liszt 1811–1886
Wagner 1813–1883
Tchaikovsky 1840–1893
Debussy 1862–1918
Strauss 1864–1949
Bartok 1881–1945
Stravinsky 1882–1971

Development of Orchestral Instruments:
Approximate starting dates of each instrument

Woodwind section:
Flute, oboe, bassoon before 1650
Clarinet 1770
Cor Anglais 1830
Piccolo 1840

Brass section:
Trumpet before 1650
Horn 1710
Trombone 1810
Tuba 1860

String section:
Violins, viola, cello
Double bass 1670

Percussion section:
Timpani before 1650
Piano 1770
Harp 1810
Celeste 1870

mation will make this a useful piece; indeed it would be most annoying not to find a composer on it, if his work was being played at the concert.

How should the starting dates of the instruments be depicted? With one known exception, these dates are all approximate. (The exception is the harp, the modern pedal-version of which was perfected in Paris in 1810 by Sebastian Erard.)

A system that graphically blurs or fudges these dates must be devised. This is a problem that often presents itself to the chartmaker—although the idea of not being exact in the depiction of facts seems to contradict the idea of the chart itself. That is sometimes why the designer must have the strength to say to the client or editor that he or she cannot make a worthwhile graphic design out of facts that are so woolly or open-ended or that have to be made to look, for political reasons perhaps, like something they are not.

In this case the blurring of the date on which the clarinet, for instance, became regularly used as an orchestral instrument is a perfectly permissable request to be made of a designer. For a start there may be differences in researched information about exactly when the date was, and then it seems only common sense to imagine that it was not an overnight event. Thus a specific date is hard to pin down. For those cases where there is obviously development, influence, and experiment—this is especially true in the arts—the chart designer must have ready a repertoire of fuzzy edges or other graphic devices needed to cope with them.

STEP FOUR

Find the Right Symbol

You should look at this part of the assignment in two ways: first to find an overall vehicle to frame and contain the information, and second for a series of small symbol illustrations to be attached to the individual instruments and perhaps portraits of the composers to go with their names and dates.

Pictorial archives are rich with engravings of music and musicians, and they will certainly provide references for the instruments and the composers. They might even be considered for use exactly as they are—in the form of engravings.

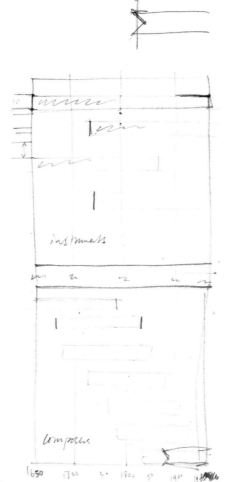

A very rough working sketch precedes the actual plotting of the data.

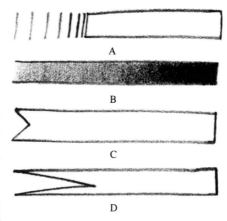

Blurring the ends of the arrows allows the "invention" dates of the instruments to be flexible within a short timespan, rather than fixing one definite year for them.

As for the overall vehicle in which to place the information, make the visual list as before.

If the concert series takes place in a particular setting, for instance, an outdoor auditorium in the country, then elements of that can also be included as a possible frame for the chart.

There is also the choice, of course, of not holding the whole design together inside an overall frame. Perhaps charting the information itself with small symbols only for the instruments is all that is needed.

SOLUTION TO THE PROBLEM

First for the composers draw out a grid of vertical lines to represent the years from 1650 to 1980 in 10-year intervals. Draw horizontal lines in this to represent the span of the composers' lives. The same format can be followed for the instruments. Armed with these two pieces, the next decision is how to put them together: Is one more important than the other? If so, which? Common to both are the years. You could begin by putting the years in the middle with the instruments at the top and the composers underneath.

Alternately, the two charts could be directly on top of one another, with the years at the bottom and a color differentiating the two sets of information.

As you do this, a general shape for the chart will appear, but you are always more or less in control of that since it depends on the thickness of, and distance between, the bars and the height of the type.

The question of how to blur the dates should be addressed at this point. Short parallel lines (A) produce a tailing-off effect, as do shaded bars (B) although these are much harder to control.

The disadvantage of (A) is the possibility of confusion arising between the grid lines and these short verticals. The disadvantage of (B) is that you need an airbrush, or you need to be able to produce this in a computer with an airbrush mode. Perhaps the best is (C), a simple arrow-back, which deliberately makes it hard to see exactly which part is meant to be read as the back, while nevertheless giving a good general idea of the time period involved. The length of the arrow-back could vary, to take into account those cases where at least some information was known about the span of the area

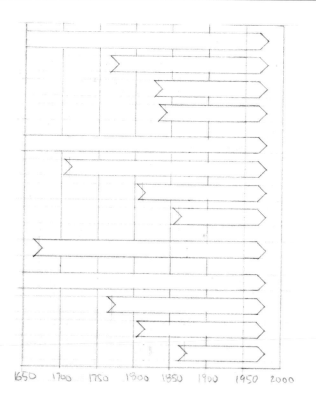

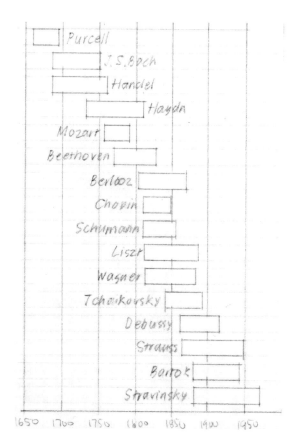

Accurate drawings of the information are placed under Mylar (see page 35) for the final artwork to be traced through.

required to be blurred. Example (D) is much less exact in its statement of the dates and therefore blurs much more than (C).

In some jobs there may be no restriction, within reason, on the amount of space the chart is to take. In this case, there is no page or column for it to conform to, as it is a separate sheet or card or insert; whether it should be folded so it can be fitted into the program is a part of the assignment that is left open to the designer. He or she can therefore let the information take its own course naturally, without forcing it into a shape defined by a page in a magazine or an advertising space in a newspaper.

This situation is good in that it could allow the symbols for the instruments to be a decent size. It would be possible to compress all the facts into a tightly fitted little space, but the bars should be spaced out somewhat if you want to make room for images of the instruments.

Two reasons will probably stop you from trying to do the same thing with the pictures of the composers that you can do with the symbols for the instruments. First, if they are reduced too much they will become impossible to recognize and will fill in—the black lines will all join together to make an ugly mess. If they are left at a readable size they will eat up too much space, detracting from the information. Second, it will be almost impossible without assigning an artist to draw them all to find a related set of pictures of each of them, and while engravings of some, reproductions of paintings of others, and photographs of the more recent ones may well provide an interesting texture, it will detract from the information by putting too much emphasis on only one part of the initial assignment.

A compromise might be to use a selection, say, half of the total number, and arrange them in such a way as to form a base for the whole design.

The final piece pulls all the elements together. Note how the composers' names cross over the top of the vertical rules and that a horizontal strip at the top and bottom clearly labels the two halves. The captioning of the pictures is done with a smaller size of the type used for the body of the chart.

Development of Orchestral Instruments

Flute, oboe, bassoon
Clarinet
Cor Anglais
Piccolo
Trumpet
Horn
Trombone
Tuba
Violins, cello, double bass
Timpani
Piano
Harp
Celeste

1650 1700 1750 1800 1850 1900 1950 2000

Purcell
J.S.Bach
Handel
Haydn
Mozart
Beethoven
Berlioz
Chopin
Schumann
Liszt
Wagner
Tchaikovsky
Debussy
Strauss
Bartok
Stravinsky

Composer's Lives

J.S.Bach

Haydn

Beethoven

Handel

Mozart

Wagner

Apartment Painting Schedule

PROBLEM

Show the progress of events in a simple apartment painting sequence.

DATA

The painters are coming on Tuesday and will complete the job by Wednesday night. The paint needs a little longer to dry. The furniture has to be moved into the center of each room. Your mother-in-law is coming to stay on Friday morning.

INTENDED USE

Personal presentation: a way of explaining to other members of the family, including children, the reasons for having to move out of the apartment for two nights.

DISCUSSION OF THE PROBLEM

There is nothing very complicated about this work-flow chart. To do a chart at all of such an ordinary event might seem to be overorganizing. But as an exercise in the preparation of this type of chart, it illustrates well the principles involved.

STEP ONE

Identify the Reader/User

Your family, and possibly friends, are the primary users of this graphic representation of what is going to be happening in your apartment. But it can also serve to let the painters themselves know that you intend to move back at a certain well-defined time and need to have the space available for an important visitor.

STEP TWO

Select a Production Method

To be really useful, several people will need copies of your efforts. You will take one to work, your spouse will keep one, your children will have cop-

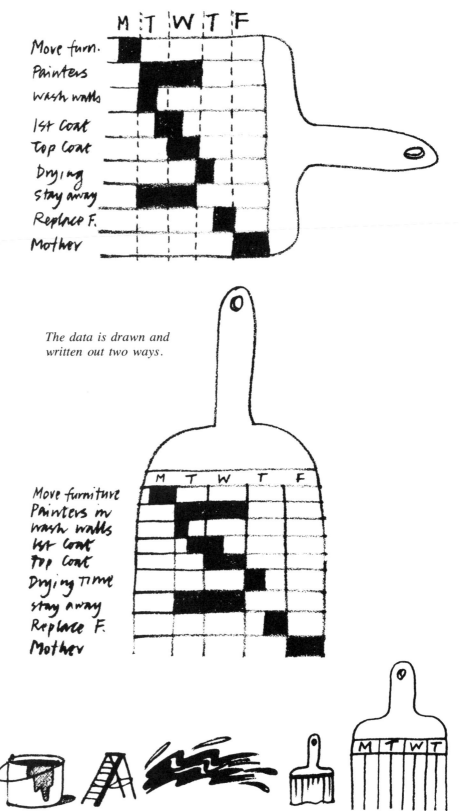

The data is drawn and written out two ways.

ies to take to school. The friends you are staying with, the people in the neighboring apartment, and the painters themselves complete the list.

Unless you are going to redo the work each time, you will probably Xerox the original in order to produce enough copies. This will restrict the size of your piece; in fact it suggests a hand-held chart, rather than a poster-sized chart. In any case since you are not in the same place all the time, there is a real value in making it portable.

Xeroxing the original usually means that it will be in one color—black—and must also be drawn to a size more or less the same as the finished Xerox. (A few copying machines can reduce and enlarge, but the majority produce same-size prints.)

Many different drawing media can be used, but simple black lines and tones made up of black lines are the most effective. A little hand coloring can be done later, using the drawing as a guideline.

STEP THREE
Review the Data
For once you are fully in charge of the amount of information to put into your chart. Do you include hours of the day or merely divisions of the day into A.M. and P.M.? Do you attempt to fill in the order in which the rooms are to be painted?

Certain facts (the painter's starting time, for instance) are clear cut and should be shown as such. But the length of time it will take to move furniture about can only be guessed, and your chart can reflect those open-ended situations.

Remembering that this is more for family consumption than professional scrutiny, keep the information to a minimum, while including all the important events.

STEP FOUR
Find the Right Symbol
Paint cans, paint brushes, splotches and splashes of paint, painters, step-ladders provide a lot of scope here. The basic form of the chart will be a series of days, with events that cross through them in the form of bars. Perhaps the most appropriate symbol—also one of the simplest to draw—is the flat paint brush. The bristles could be stylized into the divisions between and within the days.

SOLUTION OF THE PROBLEM
On a standard piece of paper (8½ × 11) make a very rough drawing of the timetable part of the chart to see what shape best suits the information.

If it is still unclear which way to draw the image is best, then proceed with the drawing of the heart of the chart: the sequence of events in the apartment. As you work ways of editing the words down to the most meaningful and briefest form will come to you.

Since you are working at a small scale, the neatest way to put the words in the chart is to type them all on one sheet of paper and then by cutting and pasting position them exactly where they should be on the drawing. The copying machine will eliminate any edges, provided you have typed on paper that matches as nearly as possible the whiteness of your base material. If you want to distinguish who is carrying out the work (that is, you or the painters), leave all the bars white and color them in after the Xerox has been made. There are few enough copies for this to be easily done, and it would certainly make the finished piece look better.

You could still show the difference in black and white by line shading or with a coarse pressure graphics screen. Or you could decide not to differentiate between activities, merely concentrating on the fact that something is going on at a given time, without the added detail of who is doing it.

Rather than a chance to simplify the information, however, it would seem to be missing a graphic opportunity not to put in that detail. At a glance everyone should know whose responsibility it is, for instance, to move furniture.

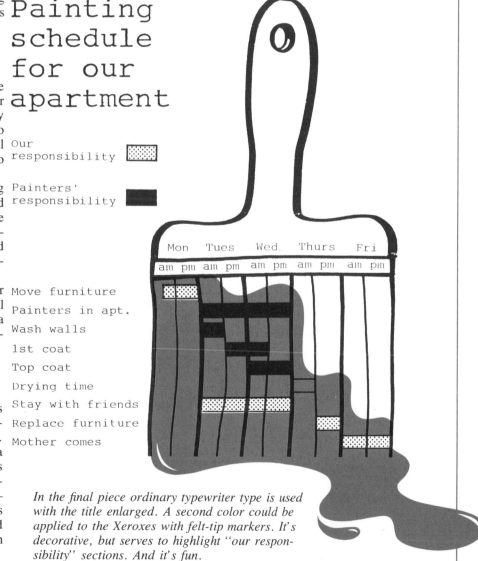

In the final piece ordinary typewriter type is used with the title enlarged. A second color could be applied to the Xeroxes with felt-tip markers. It's decorative, but serves to highlight "our responsibility" sections. And it's fun.

How the Price of Oil Increases

PROBLEM

Show the rise in the cost of oil per gallon as it travels from the ground in Saudi Arabia to two outlets in the United States: as gasoline at the pump and as domestic heating fuel.

The figures are from April 1979 at the start of a summer in which the price of gas at the pump was creeping up to the dreaded $1 per gallon.

DATA

The data you need to work with is listed in the accompanying table.

INTENDED USE

There are two ways this chart will appear in print: (1) in a daily paper in black and white in one column and (2) in a magazine in 2 color in two columns.

DISCUSSION OF THE PROBLEM

One of the facts confronting a chartmaker is that when oil is a continuing important story in the news there will inevitably be many calls for oil charts. That is of course also true of gold, stock prices, inflation, or any event that continues to be in the headlines. How to deal with this on a weekly or, worse, a daily, basis must concern the artist. A daily paper should try to set up a format that it can repeat by adding numbers easily, while a weekly magazine can afford to make more radical and imaginative changes of direction and approach to the subject.

As the price of oil began to rise in 1979, readers needed to be informed

	In $
One barrel of Saudi light crude oil	14.55
Transport to the U.S. (per barrel)	1.25
U.S. government entitlements (per barrel)	1.50
Gasoline:	**In cents**
Refining, wholesaling, marketing (per gallon)	20.4
Taxes (per gallon)	13.2
Retailer (per gallon)	9.7
TOTAL	77.3
Heating oil:	
Refining (per gallon)	14
Retail dealer and wholesaler (per gallon)	14
TOTAL	62

about the elements which added costs to something that started at only about 30 cents a barrel to pump out of the desert. Since there are 42 gallons in that barrel, the cost is less than 1 cent per gallon. What made it cost 77.3 cents at the local gas station?

STEP ONE

Identify the Reader/User

The audience, for both daily papers and magazines, has already been discussed in assignment 1, and the same holds true here.

STEP TWO

Select a Production Method

As before, the newspaper will require flat art, and the magazine flat art will need overlays for tints of the second color.

STEP THREE

Review the Data

For the purpose of this assignment, all the data has been supplied, but do not forget that in some cases you will have to spend a good deal of time researching the facts and figures for your charts. Some research is much harder than others, and this assignment is a case where you or your researcher would have had to dig very hard in a number of different sources.

The first thing to notice is that the price does not rise all the way through the data. After the "entitlements" (a way of averaging domestic and foreign oil) the price per barrel actually goes down. A way must be discovered to deal with this.

Next notice that after those entitlements, the data split into two tracks: one for gasoline, the other for heating

Refinery   Tax

oil. Also the categories are different on each track; while marketing adds something to the gasoline price, it plays no part in the heating oil total and no tax is involved there either.

Finally, the unit of measurement changes en route from barrels to gallons. There are good reasons for keeping that: Crude oil is always counted in barrels, whereas the gallon is a far more understandable unit at the other end of the chart where it affects the consumer. This difference, however, must be made clear typographically.

STEP FOUR
Find the Right Symbol
The symbols you choose need to be closely tied to the type of media the particular chart will appear in.

1. Newspaper: Within a restricted space, such as that available here, particular attention should be given to the typography. This chart has more than the usual labeling and must be allowed enough space to be read clearly. Therefore the symbols or illustrations for the newspaper solution should be kept rather minimal—one to represent the gas pump and one to represent home heating oil.

2. Magazine: In the more generous space afforded by a magazine, it may be possible to illustrate all the steps in the progression suggested by the data.

SOLUTION TO THE PROBLEM
Because of the complexity of the data in this assignment, the following discussion is quite detailed.

Newspaper: In the interests of keeping the information simple, the chart should start with the point at which the oil has arrived at the refinery. This eliminates the difficulty involved with the lowering of the price from its initial figure. You must of course be prepared with sketches that do include all the information from the original data list, but they should be a good enough argument in themselves to convince an editor of the wisdom of simplification in this case where space is limited.

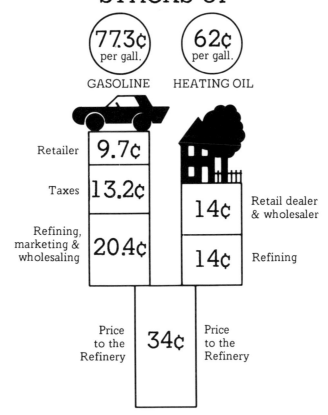

HOW THE PRICE STACKS UP

Caption: *The straightforward newspaper usage is clean and unfussy. Note that only one sort of type is used in various sizes.*

Simplification is good as long as you understand exactly what you are doing: What you gain in clarity must be weighed against the loss of some information. Occasionally you may be rewarded with the extra space needed to allow you to chart all the data.

This simpler approach to the assignment means that only one part of the chart is common to both gasoline and heating oil: the cost to the refinery in the United States.* Use this as the base of the chart and build two bars from it.

The amounts can actually be charted—that is, measured and shown in

*It also means that the whole chart can be measured in gallons. The price at the refinery is $14.30, $14.55 official price, plus $1.25 for transport, then minus $1.50 entitlements, thus equaling $14.30. At 42 gallons per barrel, that equals 34¢.

proportion to one another—so proceed up the two bars until all the figures have been dealt with and put the two symbols, for gas and home, at the top.

The bars should be only as wide as is necessary to accommodate the numbers; the descriptions should be placed outside the bars. This will give you much more flexibility to stack the type in a variety of ways.

Note how the bars suggest a title for the chart: "How the Price Stacks Up." This has more life than the original given title, which should always be thought of as a starting or working title. You may be surprised at how often you will think up the best title for your chart and how easily that will be accepted by editors or clients.

Magazine: The aim here is to make a design out of the elements. By piling all these parts one on top of each other, the impression will be given of a large increase stemming from a small beginning. To add graphic and illustrative interest a dromedary is used at the base to symbolize the Arabian desert. Make sure that you do your homework about this sort of detail.

The animal that is found in Saudi Arabia is the one hump (*Camelus Dromedarius*), not two (*Camelus Bactrianus*). Although the barrel—the next element in the illustration—

Make sure your zoology is correct— on the left, a camel, on the right, a dromedary!

would fit neatly between the two humps of the bactrian camel, it would be an embarrassing convenience!

The outstretched hand of Uncle Sam

has been replaced by his hat. It forms a better pedestal for the gas pump and the oil tanker and avoids the necessity of showing a truncated arm and hand in the middle of the chart. By leaving the hat uncolored on the heating oil side, the fact that no tax is added there is visually underlined.

The entitlements scheme of averaging out domestic and foreign crude oil (really a refund from the government, reducing the cost of foreign oil to the U.S. refineries) is shown as a pair of scales, with the balance tipped on the foreign side.

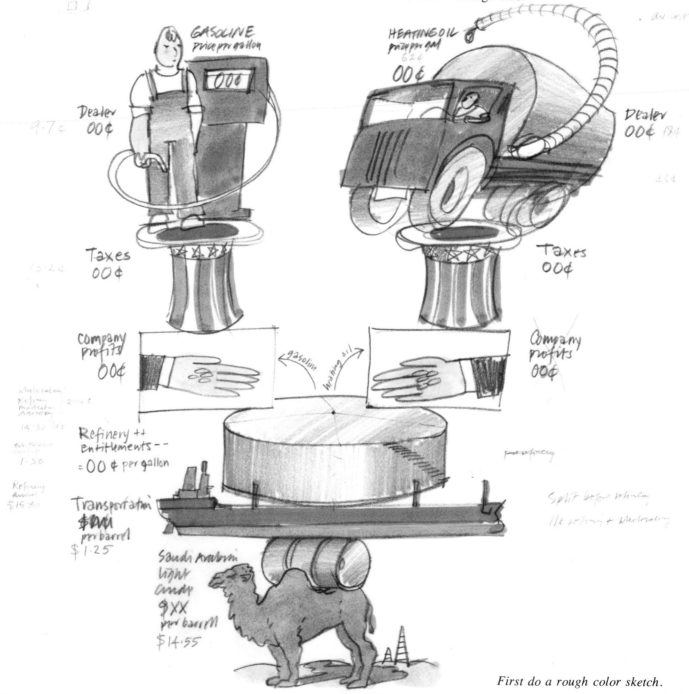

First do a rough color sketch.

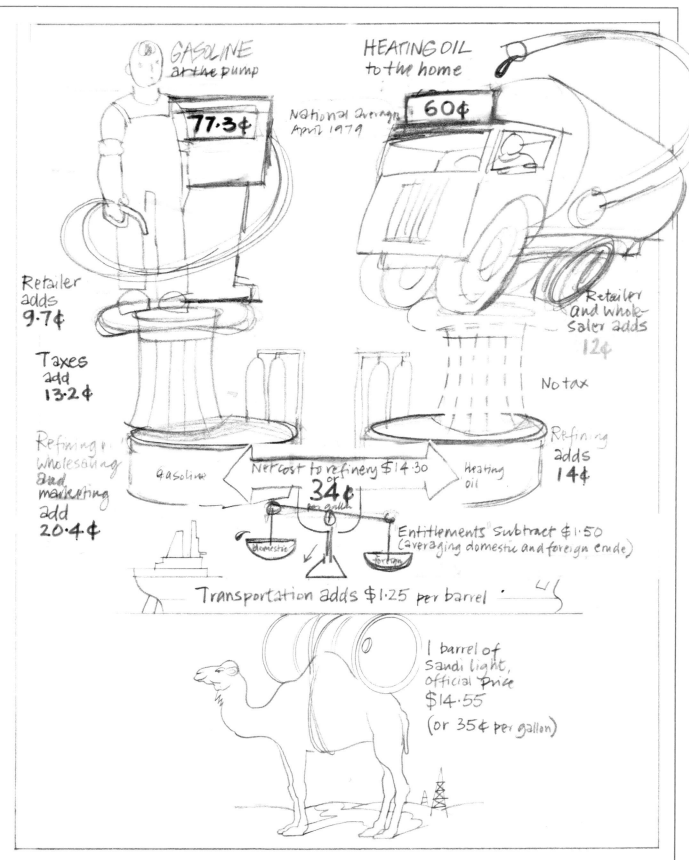

GASOLINE
at the pump

HEATING OIL
to the home

77.3¢

National average
April 1979

60¢

Retailer
adds
9.7¢

Retailer
and whole-
saler adds
12¢

Taxes
add
13.2¢

No tax

Refining,
wholesaling
and
marketing
add
20.4¢

Gasoline

Net cost to refinery $14.30
or
34¢
per gallon

Heating
oil

Refining
adds
14¢

Entitlements subtract $1.50
(averaging domestic and foreign crude)

Transportation adds $1.25 per barrel

1 barrel of
Saudi light,
official price
$14.55

(or 35¢ per gallon)

The second color sketch shows the elements in their final agreed-upon form. Note the changes between these two stages. This is where all the facts must be ironed out, not later at the artwork stage.

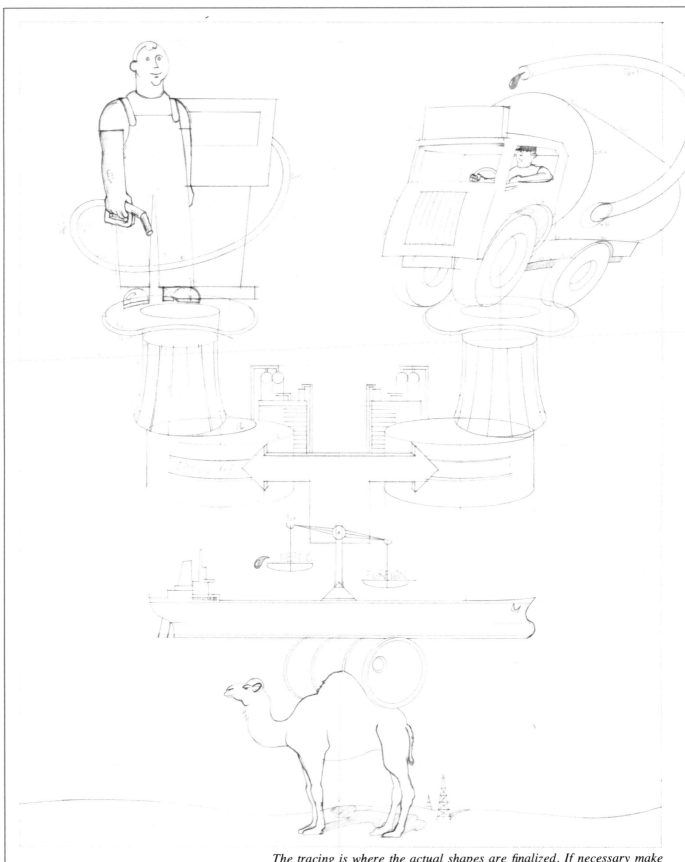

The tracing is where the actual shapes are finalized. If necessary make notes to remind yourself of the curves and templates used to create the lines. This annotation will speed up the production of the final art and ensure that no mistakes are made when, with the deadline approaching, time will probably be getting tight.

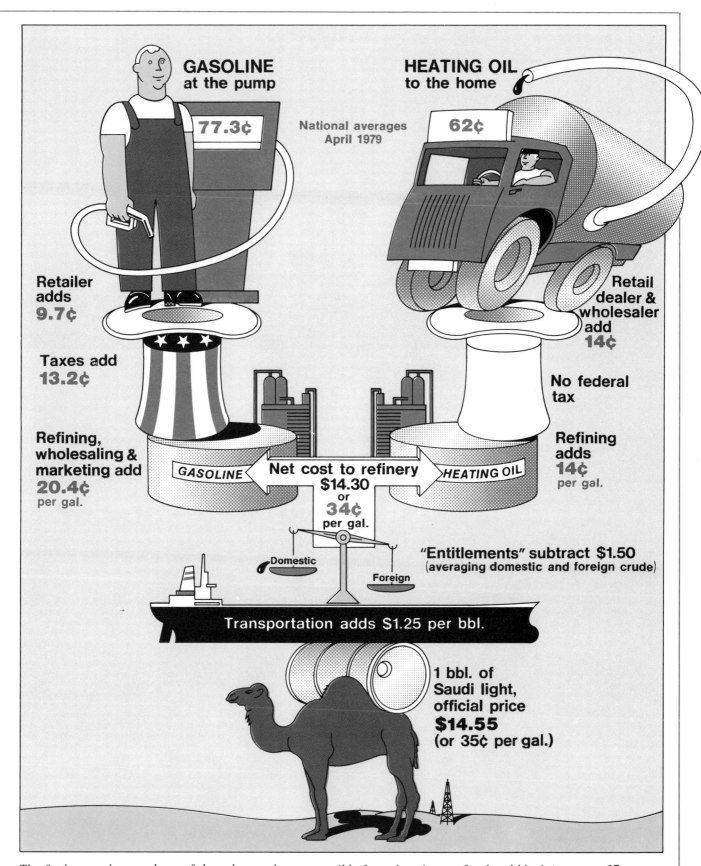

GASOLINE at the pump

77.3¢

National averages April 1979

HEATING OIL to the home

62¢

Retailer adds **9.7¢**

Retail dealer & wholesaler add **14¢**

Taxes add **13.2¢**

No federal tax

Refining, wholesaling & marketing add **20.4¢** per gal.

GASOLINE

Net cost to refinery **$14.30** or **34¢** per gal.

HEATING OIL

Refining adds **14¢** per gal.

Domestic

Foreign

"Entitlements" subtract $1.50 (averaging domestic and foreign crude)

Transportation adds $1.25 per bbl.

1 bbl. of Saudi light, official price **$14.55** (or 35¢ per gal.)

The final art makes good use of the colors and tones possible from the mixture of red and black (see page 37).

Company Management Diagram

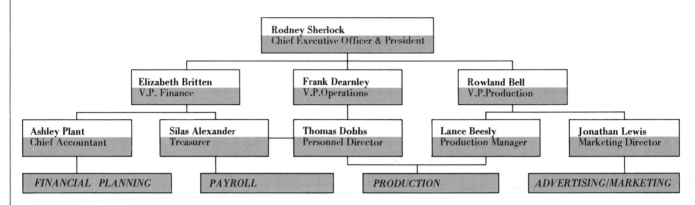

In an annual report the result is a no-nonsense, businesslike table.

PROBLEM

Show the hierarchy and reporting responsibilities in a typical manufacturing company.

DATA

The Chief Executive Officer and President of the company is Rodney Sherlock. Top management is Elizabeth Britten, Vice-President, Finance; Frank Dearnley, Vice-President, Operations; Rowland Bell, Vice-President, Production. Reporting to Elizabeth Britten are Ashley Plant, Chief Accountant, and Silas Alexander, Treasurer. Reporting to Frank Dearnley is Thomas Dobbs, Personnel Director; to Rowland Bell are Lance Beesly, Manager, Production, and Jonathan Lewis, Marketing Director. The chart should end at the four broad areas under the last rank mentioned: Financial Planning, Payroll, Production, and Advertising/Marketing.

INTENDED USE

This chart will be used (1) in the company's Annual Report and (2) as a grade-school teaching tool.

DISCUSSION OF THE PROBLEM

Since the officers of the company require a very strict adherence to the ranking of their names and jobs, the presentation of this information in their annual report would, at the outset, appear to be a rather cold prospect. Annual reports are notoriously difficult, since the personalities of so many individuals are involved. Decoration or illustration in this type of work is largely shunned by all, save the most self-assured or relaxed companies. It may therefore be better to curb one's natural desire to illustrate the various functions/jobs involved and instead include simple head shots of the people named. It is indeed a pity that such companies take themselves so seriously, but in general their argument has to do with presenting a sober image of themselves to their shareholders. Humor has not yet attained the degree of respectability that it should have.

The story will be different in the schoolroom situation, where a more lively approach can be attempted.

STEP ONE

Identify the Reader/User

Here the differences between readers couldn't be greater.

1. Annual Report: Whoever the readers of the report may actually be, the clients are the people whose names are in it. Thus the instincts of the designer about what would really make an interesting chart must this time be suppressed in deference to the sometimes philistine desires of those who are paying the bill. If this seems an overly pessimistic view of the situation, it is merely the result of experience. Anyway, as long as you are prepared for the worst, it can only get better! And besides none of these warnings should stop you from trying. Sooner or later the barriers will come down; some already have, as seen in Chapter 5.

2. Classroom: Children need to be taught first about the concept of abstract diagrams representing real things. By using drawings of people doing things, instead of boxes simply containing their names and job titles, young imaginations can be helped to understand and remember.

STEP TWO

Select a Production Method

Because of the difference in the readership you can choose a very formal, precise method for one and a very informal one for the other.

1. Annual Report: The chart will be produced by drawing it as flat art with overlays to indicate color. Annual reports are invariably printed on very good quality paper, and a great deal of trouble is taken over the platemaking and production, since this becomes the company's ultimate advertisement. You should take care with the preparation of every piece of art, but this type of work requires extra special attention to detail.

2. Classroom: This chart is to be drawn on a blackboard by the teacher in front of the children. As such it differs from all the other solutions to assignments in this chapter in that it unfolds before the eyes of the audience and is accompanied by a spoken commentary as the drawing is completed.

STEP THREE

Review the Data

As previously noted, the limits are clearly defined here. All that must be done is to present the hierarchy as you received it. A pyramid shape is immediately suggested by the number of people and their rank: one man at the top, a board of four directors under him, and then a third rank of five under the directors. At the foot of the chart are four broad areas that are not detailed, but form a convenient base for the diagram.

STEP FOUR

Find the Right Symbol

What "symbol" is most appropriate, considering the different solutions to the problem?

1. Annual Report: In the light of the previous discussion, "symbol" is perhaps the wrong word here. The chart, however, must have a "form" in which to appear. Neatly ruled boxes carefully designed to accommodate the type and then joined elegantly by lines to denote reporting responsibilities should be used. A typographic difference can be made between name and job title. Placing the chart within the total design of the page should also be considered.

2. Classroom: Symbols for each of the people could be worked out, but since this is in the nature of a live performance piece, it might be simpler and quicker to limit the symbols to a stick figure of some sort for each person mentioned and to use different symbolic drawings only for the four broad areas at the foot of the chart. Do not forget the woman is a different shape!

SOLUTION OF THE PROBLEM

One solution requires much more attention to detail than the other.

1. Annual Report: Draw out the boxes in the order that they appear. Specify the correct size for the type: larger for the names than for their job descriptions. When you have the type, check that the boxes are the correct size. To look elegant and intentional, the type must fit within the boxes properly,

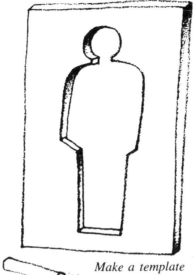
Make a template out of cardboard to help you draw the characters easily.

with neither too much space, nor too little around them.

For an annual report, the names are more important than the jobs, so put a tint over the part of the box containing the job description. Make a balanced design of all the parts.

2. Classroom: Here the names of the officers of the company are far less important than their jobs within the organization. Therefore in this solution the names will be omitted and replaced simply with the symbols for man or woman. Their job titles remain.

Practice drawing out the boxes and their attendant symbols, leaving a few key marks on the blackboard as a guide for the final drawing in the class. You might consider making a template as a guide for the human figures. It would not take long to cut it out of heavy card, and it would ensure a uniform size for the "people" on the chart, with a minimum of effort and concentration—at a time when some of those senses should be engaged in watching the students! Keep the illustrations at the base of the chart as simple as possible. Contain them all within the same shape to keep them equal weight.

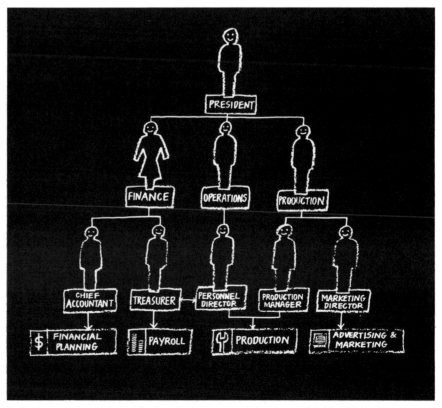
Talk the students through the chart as you draw it on the blackboard.

Airline Timetable

PROBLEM

To promote one particular route, a large airline has decided to single that out from its normal comprehensive airline timetable for special treatment.

DATA

The route is between Los Angeles and the three New York airports—Kennedy (JFK), La Guardia (LG), Newark (N). Also to be included are the frequency of the flights (do they fly every day?), the availability of meals, whether or not they are nonstop or direct flights (that is, with stops but not changing plane), flight numbers, type of aircraft, when night coach fares apply.

Flights Leave NY	Arrive LA
8:30 am (LG)	1:10 pm
9:30 am (N)	2:10 pm
11:00 am (JFK)	1:50 pm
4:30 pm (JFK)	7:25 pm
6:45 pm (JFK)	9:45 pm
7:45 pm (LG)	11:45 pm

Leave LA	Arrive NY
7:30 am	5:30 pm (LG)
8:30 am	4:30 pm (JFK)
10:00 am	6:00 pm (JFK)
11:00 am	8:30 pm (LG)
1:15 pm	10:45 pm (LG)
3:30 pm	11:20 pm (JFK)
10:00 pm	5:45 am (JFK)
12:45 am	10:24 am (LG)

INTENDED USE

It is an additional piece of printed information available at travel agencies, airline offices, airports; it is possible that later the material will be translated into an ad for newspapers.

DISCUSSION OF THE PROBLEM

Competition is strong on this route across the United States. The public is baffled by the constant price wars between the airlines. This chart is to present the simple facts of when there are flights for this particular airline.

STEP ONE

Identify the Reader/User

Airlines make a great deal of their money from business travelers, and since they have to fly regularly they are, in a sense, a captive audience, less likely to need a special timetable. After all, this information can all be found in the airline's regular printed schedule; and it is quite likely that the businessperson's company will make the reservations, anyway.

Therefore, the main reader or user of this timetable is the more casual, perhaps less experienced, traveler. Possibly it is someone going on vacation. Certainly they may not be used to reading the densely packed type of the worldwide flights of one airline. Therefore this assignment is as much an exercise in advertising the schedule as it is in printing the information. A way must be found to make it easy to understand and engaging to look at.

STEP TWO

Select a Production Method

The chart is to be printed in 2 color and must conform to the size of the airline's regular timetable, which is 4 by 9 inches. It is to be a single sheet, printed on one side only.

Flat art with overlays for notes and for the second color will be the best production method.

STEP THREE

Review the Data

A great deal of information has to be included apart from the flying times. Some of this is less important—like the type of aircraft or, at this stage, the flight number—but nothing can be omitted if it is to be a really useful guide for the prospective flyer.

A system of columns will have to be worked out to accommodate the nine types of information:

1 Flight departure time
2 Departure airport
3 Flight frequency
4 Meals
5 Nonstop/Direct flight
6 Flight number
7 Type of aircraft
8 Fare type
9 Flight arrival time

In one case it should be possible to eliminate the separate column: The nightcoach fare only applies to four of the fourteen flights. An asterisk can

take care of that. An analysis of the type of aircraft also reveals that while a number of different aircraft are flown on this route, they can be simplified into widebody (for example, 747s) and others (for example, 707s). This basic difference is all the average passenger needs to know about the equipment being used. This difference could be indicated by printing all the widebody flights in bold type and the others in normal face.

STEP FOUR
Find the Right Symbol
As before, start with a roughly drawn "shopping list" of likely images that go with flying and with New York and Los Angeles. Check whether anyone would object to a "landscape" rather than a "portrait" (or upright) use of the area. The regular schedule is in fact upright. If it's OK to use the landscape shape you can use it to advantage.

SOLUTION TO THE PROBLEM
Write out all the information that has to be included. This will give you a good idea of the "shape" of the information and how long each line of information will be for each flight.

Note the abbreviations that can automatically be used:

ex = leaves
NY = New York
8:30a = 8:30 am
1:10p = 1:10 pm
LAG = La Guardia
JFK = J.F. Kennedy
EWR = Newark
LAX = Los Angeles

(These are the official International Airport abbreviations.)

Frequency:
M = Monday
Tu = Tuesday
W = Wednesday
Th = Thursday
F = Friday
Sa = Saturday
Su = Sunday
X = not on a day stated
D = Daily
M = Main Meal
S = snack
Stops: 0 = none, etc.
Arr = arrives

ex NY	Airport	Freq	Meal	Stops	Flt#	arr LA
8.30a	LAG	XF	M	1	102	1.10p
→9.30a	EWR	D	M	1	212	2.10p
11.0a	JFK	D	S	0	213	1.50p
→4.30p	JFK	XSu	M	1	578	7.25p
→6.45p	JFK	D	M	0	214	9.45p
7.45p	LAG	D	M	1	579	11.45p

ex LA	Freq	Meal	stops	Flt#	arr NY	airport
7.30a	XSa	M,S	1	101	5.30p	LAG
→8.30a	D	M	0	211	4.30p	JFK
→10.0a	D	M	0	214	6.0p	JFK
→11.0a	XSu	M,S	1	577	8.30p	LAG
1.15p	XM	M,S	1	314	10.45	LAG
3.30p	D	M	0	103	11.20	JFK
10.0p	D	M	0	104	5.45a	JFK
12.45p	XSa	S,M	1	580	10.24a	LAG

→ Bold Face 747
rest 707

Writing out everything that has to be set in type helps you understand the information, and it will show you what shape is the most natural for it.

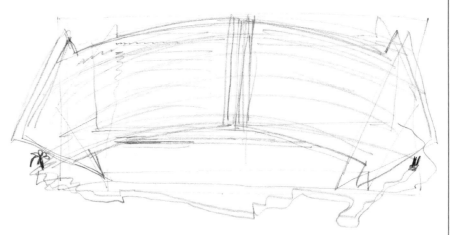

Roughly draw out the arrow idea to confirm whether or not it's going to work.

Unless a very condensed type is being used, the table for flights from New York to Los Angeles is going to come out wider than it is deep. In the case of Los Angeles to New York writing it all out will show you how the table will appear.

Note that all flights from Los Angeles leave from the same airport, so the airport column is redundant there, but it must be placed near "arr. NY." You will note that this is also a horizontal shape, even though there are two more flights in this direction.

Returning to the dimensions you have to work within, it seems that a natural shape for a graphic representation of flights between the East and West Coasts is the horizontal or landscape shape. If you put the shape of the United States on this, tilted back somewhat to flatten it, will there still be room for the tabular matter? Try it.

As with most of these problems, there is a constant switching from illustration idea to information and back again, until a graphic harmony is reached between the two.

To underline what has already been said: The reason for the existence of the chart is the desire to communicate the information. The illustration must not dominate, but amplify.

A part of this back-and-forth procedure will be to set the information in type. Perhaps, if the budget allows, you will try two different faces—one condensed and one regular (with bold). In any case it is good practice to trace very accurately from a type specimen at least one entire line, so that you can specify correctly.

Of the original stock of possible symbols, the symbolic representations of the two cities are omitted, as they finally appeared to be too decorative and fussy. The arched, double-ended arrow flowing from Los Angeles to New York and vice-versa well serves as the vehicle for the type and the idea of flying between two points.

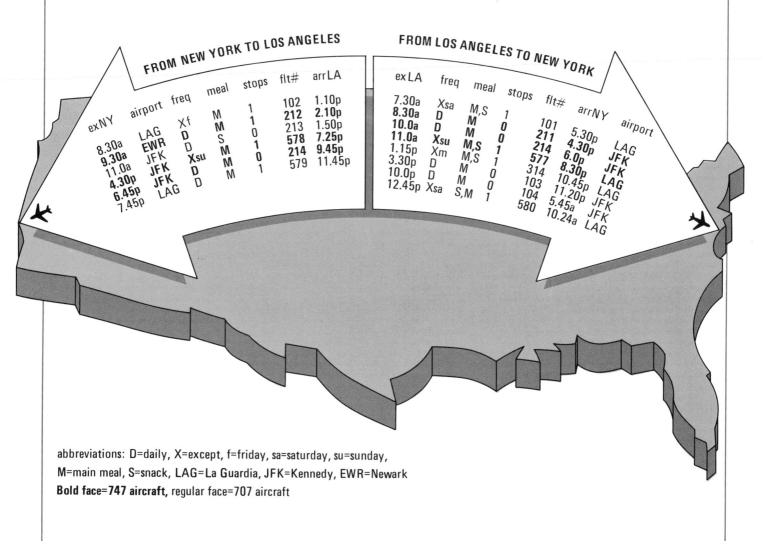

abbreviations: D=daily, X=except, f=friday, sa=saturday, su=sunday,
M=main meal, S=snack, LAG=La Guardia, JFK=Kennedy, EWR=Newark
Bold face=747 aircraft, regular face=707 aircraft

In the final art the type is curved to follow the shape of the arrows, and a tiny silhouette of an aircraft is placed at the very tip of each point.

Duty Roster at a Summer Day Camp

		MONDAY	TUESDAY	WED.	THURSDAY	FRIDAY
🕚	11·0 am Snack	Barb/Rob John/Mary	Barb/Billy	Rob/Susan	Billy/Mary	John/Susan
🕧	12·30 Lunch	Billy	Robert	Susan	Mary	Barbara
🕜	1·30 Washing up	Barb/John	Billy/Susan	John/Mary	Rob/Barb	Rob/Susan
🕞	3·30 Snack	Susan	John	Billy	Barbara	Mary
🕕	6·0 Preparation	Billy	Mary	Robert	John	—

As with the previous assignment, writing out all the information is a necessary step in understanding how best to display it. This is one of the three ways the data can be ordered.

PROBLEM

Display the names of six children in such a way that they will understand what jobs they are assigned during the five periods of the day, all of which are in some way connected to meals or snacks.

DATA

The information you need are the children's names—Barbara, Billy, John, Mary, Robert, Susan—the days—Monday through Friday—and the duty times—11 am snack, 12:30 lunch preparation, 1:30 washing up, 3:30 afternoon snack, 6:00 preparation for tomorrow (except Friday).

INTENDED USE

The duty roster will be posted on the camp bulletin board.

DISCUSSION OF THE PROBLEM

This is a two-part assignment. First the children must be equitably distributed among the jobs in the week, with some of the tasks needing two children and others only one.

Second, the duties must be presented in a timetable form that is quickly understood by children, whose thoughts may well be on other less mundane things than preparing a meal or washing up after it!

STEP ONE

Identify the User

Children aged 9–15 are the most important users of this timetable. Camp counselors will also refer to it.

STEP TWO

Select a Production Method

As a one-time chart, this will be drawn by hand on a conveniently sized piece of board or paper. The lettering can be applied with a stencil or by pressure-sensitive graphic letters, or you can type it on a typewriter.

STEP THREE

Review the Data

There may well be some mathematical formula or a computer program that will fit the number of children into the number of jobs, taking into account those which need two people and those which need only one. Most people, however, will use the simpler method of trial and error to make the various duties fit into the week's grid. Whichever method you use, only you can decide whether a fair distribution has been made.

For the second part of the assignment, you must decide which of three ways to present the timetable is the best. Each distinctly different way highlights a different fact. Depending on how you arrange the information, the children, the job, or the days will be shown in the most important light. When reviewing the data, you should decide which of the three is the most important.

STEP FOUR

Find the Right Symbol

As the announcer of some rather unwelcome but necessary information this chart could well do with some

livening up to lighten its bad news. Clock faces could denote the times of the jobs. Symbolic drawings could represent the jobs themselves. Knives, forks, and spoons could be used as dividers between the columns of information. The whole chart could be set inside the "frame" of an apron.

SOLUTION TO THE PROBLEM
Trial and error will take its own course. Once a fair distribution of the jobs is achieved, the three different ways of presenting the same information can be filled in. Next, the symbols can be put into position.

Although it would be possible to work out the positioning of all the type and produce the whole chart on a typewriter, this might be too small for the intended purpose. Careful hand-lettering or stenciling would be much better.

Many different styles of lettering are available as stencils, but you must practice with them before doing the final piece. If you are in a hurry do not attempt to use stencils. They will look less good than an honest hand-lettered approach. Although they never replace typesetting, stencils do produce letters of a uniform size and shape, and therefore they must be properly aligned and spaced to look effective.

"Instant" lettering (pressure graphics) are a possibility, especially for the second and third examples, where there is less actual lettering to be done (the names, for instance, are only spelled out once).

Depending on the time at your disposal, and your desire to entertain at the same time as inform, there is one more way of completing this assignment. If you buy a plastic-coated, plain-colored (preferably white) kitchen apron, it could be pinned to the bulletin board—or even have its own board—and used as a surface for the entire graphic presentation. There are a number of markers and pencils on the market that will draw on plastic and glass. Some will wipe off, some will not. Obviously, if you need to change the information (or if you want the apron at the end of the summer!), use the wash-off type.

Once your chartmaking has reached this stage of relaxed free thinking, consider finally the idea of not pinning it up on a board, but of appointing the cook the "task" of wearing the chart!

DUTY TIMES

	MONDAY	TUESDAY	WED.	THURSDAY	FRIDAY
BILLY	12·30/6·0	11·0/1·30	3·30	11·0	—
SUSAN	3·30	1·30	12·30	11·0	11·0/1·30
JOHN	11·0/1·30	11·0/3·30	1·30	6·0	11·0
MARY	11·0	6·0	1·30	11·0/12·30	3·30
ROBERT	11·0	12·30	6·0	11·0/1·30	1·30
BARBARA	11·0/1·30	11·0	—	1·30/3·30	12·30

	11·0 Morn Snack	12·30 Lunch	1·30 Washup	3·30 Snack	6·0 prep
BILLY	Tu, Th	M	Tu	W	M
SUSAN	W, F	W	Tu, F	M	—
JOHN	M, F	—	M, W	Tu	Th
MARY	M, Th	Th	W	F	Tu
ROBERT	M, W	Tu	Th, F	—	W
BARB	M, Tu	F	M, Th	Th	—

Here are two other ways of listing the information (the first is on page 111).

For the final piece, it was decided to leave out the symbols—there was enough going on to entertain the eye. The second of the three ways of presenting the duties was selected as being the easiest of the three to read. Let's hope the cook was good humored!

DUTIES

	11:0 Snack	12:30 Lunch	1:30 Washing up	3:30 Snack	6:0 Prep
BILLY	Tu, Th	M	Tu	W	M
SUE	W, F	W	Tu, F	M	—
JOHN	M, F	—	M, W	T	Th
MARY	M, Th	Th	W	F	Tu
ROB	M, W	Tu	Th, F	—	W
BARB	M, Tu	F	M, Th	Th	—

Chapter Five
Good Examples

FOR A BREAK FROM all the how-to-do-it and the preaching about how-not-to, this chapter presents a selection of fevers, bars, pies, and tables, all chosen for one good reason or another.

While every chart should dispense information, some are more successful at their job than others. This may be for a variety of reasons. Elegance, simplicity, humor, clarity, and responsibility can all be employed by the designer of charts, as by the designers of advertisements, magazine layouts, editorial illustrations, or any graphic matter. They are all ways of making the reader take notice of the design. If you can get the attention of your reader, you are half way to helping him or her understand and remember the facts.

The examples on the following pages speak for themselves. They are variously elegant, simple, funny, clear, and responsible and as such are good examples from a field that is growing as fast as editors and clients realize the potential that has been there since the first charts were made by Playfair in 1800.

They are grouped according to the categories already established in this book: first fever charts, then bars, then pies, and finally tables. Among the sources scanned for this brief survey are magazines, newspapers, annual reports, advertisements, and books.

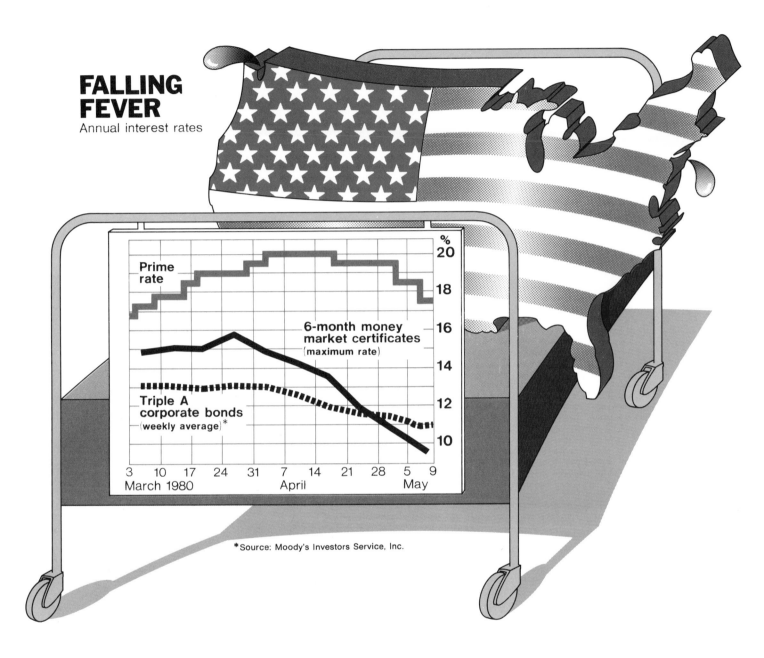

FALLING
FEVER
Annual interest rates

Prime
rate

6-month money
market certificates
(maximum rate)

Triple A
corporate bonds
(weekly average)*

%
20
18
16
14
12
10

3 10 17 24 31 7 14 21 28 5 9
March 1980   April   May

*Source: Moody's Investors Service, Inc.

This chart plays on the origin of the name for this type of chart. Is the patient getting better as its fever falls? Note the use of a rather Victorian style bed. Nowadays hospital beds no longer look like our idea of a bed. However, this is what a bed looks like symbolically, and it would have been unnecessarily confusing to draw the kind actually in use today. Its odd appearance would have attracted too much attention away from the information.

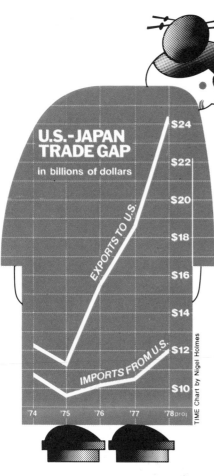

U.S.-JAPAN TRADE GAP

in billions of dollars

EXPORTS TO U.S.

IMPORTS FROM U.S.

$24
$22
$20
$18
$16
$14
$12
$10

'74 '75 '76 '77 '78 proj

TIME Chart by Nigel Holmes

Here are two charts on the same subject. (Right) A reference to Japanese woodblock prints, with a good "movable" element in the warrior's arm, allows the line to go where it should and not be forced into a shape that is unnatural for the drawing, nor the wrong shape to accommodate the information. (Above) The flat back of the kimono provides a large empty space to contain the figures. In both instances the superiority of Japan's trade situation is reflected in the image—one aggressive, the other a knowing, slightly smug over-the-shoulder look.

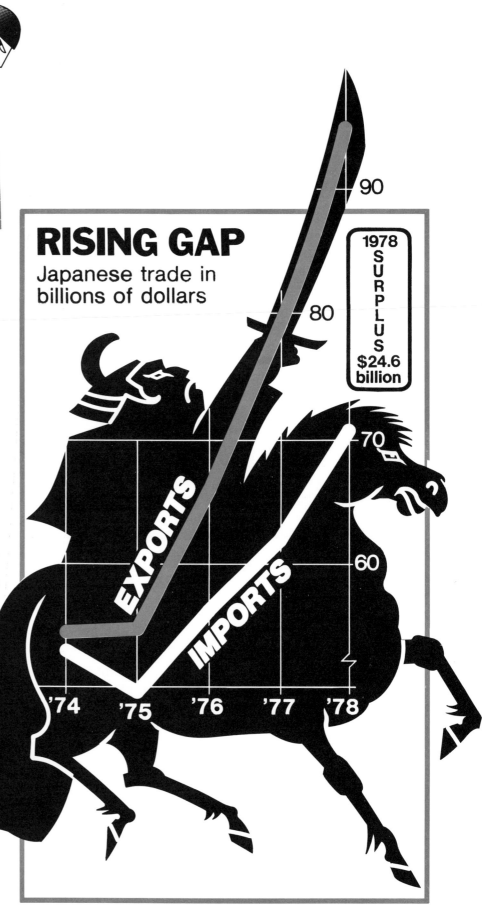

RISING GAP

Japanese trade in billions of dollars

1978 SURPLUS $24.6 billion

EXPORTS

IMPORTS

90
80
70
60

'74 '75 '76 '77 '78

The decline in EPA spending

(EPA grants to state, local governments in millions of dollars)

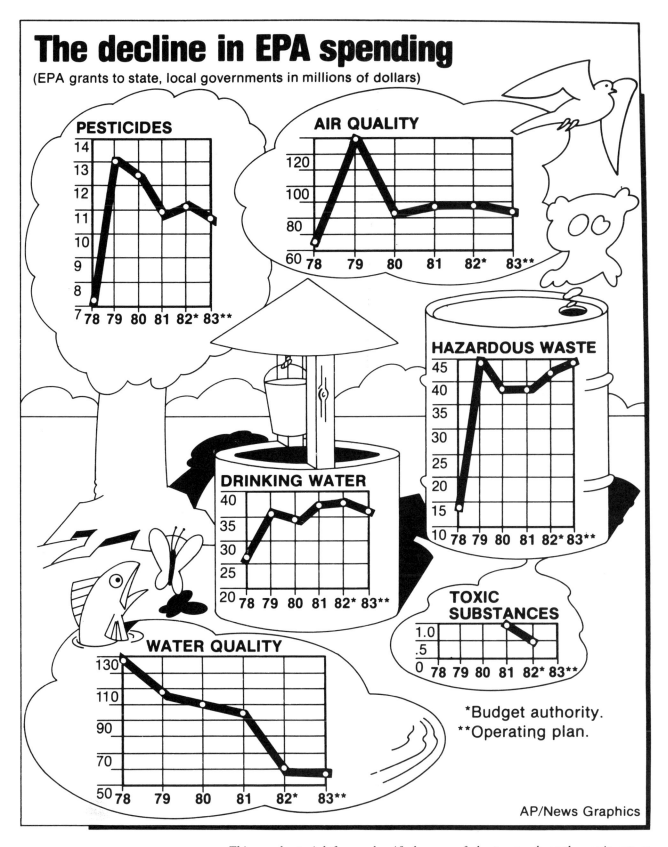

PESTICIDES

AIR QUALITY

HAZARDOUS WASTE

DRINKING WATER

TOXIC SUBSTANCES

WATER QUALITY

*Budget authority.
**Operating plan.

AP/News Graphics

This good, straightforward unified group of charts are about the environment. Just the right amount of light, decorative surrounding drawing with clear fever lines is perfect for its intended use in black-and-white newspapers. Although the sales are clearly marked, an indication that they do not start at zero would have been preferable.

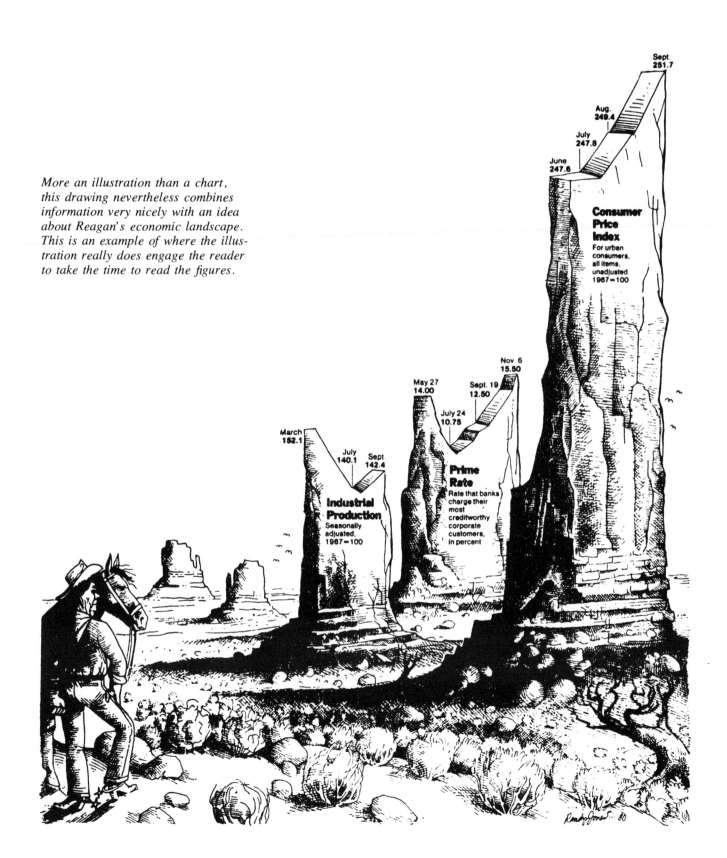

More an illustration than a chart, this drawing nevertheless combines information very nicely with an idea about Reagan's economic landscape. This is an example of where the illustration really does engage the reader to take the time to read the figures.

Consumer Price Index
For urban consumers, all items, unadjusted 1967=100

Sept. 251.7
Aug. 249.4
July 247.8
June 247.6

Industrial Production
Seasonally adjusted, 1967=100

March 152.1
July 140.1
Sept. 142.4

Prime Rate
Rate that banks charge their most creditworthy corporate customers, in percent

May 27 14.00
July 24 10.75
Sept. 19 12.50
Nov. 6 15.50

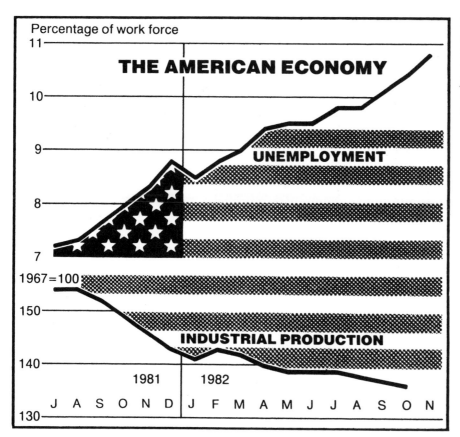

THE AMERICAN ECONOMY

Percentage of work force

UNEMPLOYMENT

INDUSTRIAL PRODUCTION

1967 = 100

1981   1982

J A S O N D | J F M A M J J A S O N

This newspaper chart uses a very good idea to link together unemployment and industrial production in America. Its obvious simplicity is all the more effective for publication outside the United States, where the stars and stripes are not used so often.

Ron Pickering says: ' The sprinter's speed comes from the force with which his feet are applied to the ground. At 60 metres, when maximum speed is reached, the sprinter's feet are no longer in contact with the ground long enough to apply the force required. Therefore, he must decelerate. The Russian **Valeri Borzov's** " overdrive " is the remarkable capacity to maintain his speed longer than any other sprinter. It's an optical illusion, but, because he is maintaining his speed while the others are decelerating. Borzov actually appears to be going faster in the sprint's closing stages '

Borzov in overdrive

other sprinters

Feet moving so fast – minimum force

Feet in maximum force position

Speed (metres per sec)

(10m) (20m) (30m) (40m) (50m) (60m) (70m) (80m) (90m) (100m)

Distance (metres)

The addition of two little line drawings on the subject of the fever line lifts this chart out of the too-straightforward approach, without interfering with the data.

119

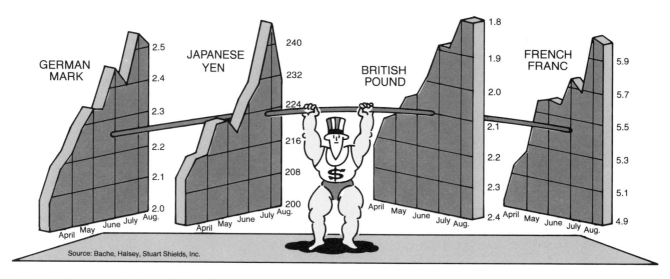

Source: Bache, Halsey, Stuart Shields, Inc.

(Above) The strength of the U.S. dollar against four other currencies is cleverly shown in this example of three-dimensional fever charts.

(Below) It is perhaps a little difficult to read some of the information set along the rooflines in this set of fever charts about employment in Kansas City, but who says you won't take the time if the drawing is enticing enough?

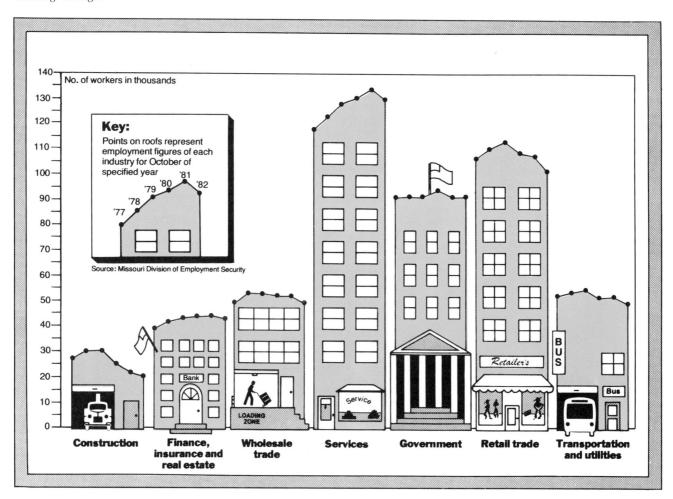

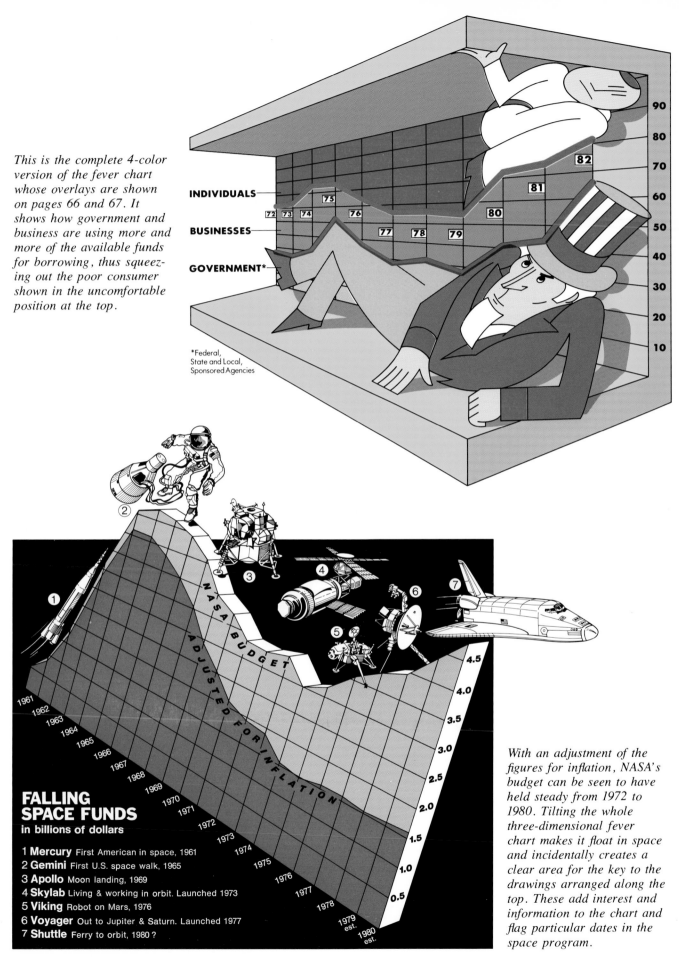

This is the complete 4-color version of the fever chart whose overlays are shown on pages 66 and 67. It shows how government and business are using more and more of the available funds for borrowing, thus squeezing out the poor consumer shown in the uncomfortable position at the top.

INDIVIDUALS

BUSINESSES

GOVERNMENT*

*Federal,
State and Local,
Sponsored Agencies

FALLING SPACE FUNDS
in billions of dollars

1 **Mercury** First American in space, 1961
2 **Gemini** First U.S. space walk, 1965
3 **Apollo** Moon landing, 1969
4 **Skylab** Living & working in orbit. Launched 1973
5 **Viking** Robot on Mars, 1976
6 **Voyager** Out to Jupiter & Saturn. Launched 1977
7 **Shuttle** Ferry to orbit, 1980 ?

NASA BUDGET ADJUSTED FOR INFLATION

With an adjustment of the figures for inflation, NASA's budget can be seen to have held steady from 1972 to 1980. Tilting the whole three-dimensional fever chart makes it float in space and incidentally creates a clear area for the key to the drawings arranged along the top. These add interest and information to the chart and flag particular dates in the space program.

Was Paul Volcker to blame for high interest rates in 1982? Whether he was or not, he was the popular scapegoat. In this large chart, reproduced from the actual two-page spread (shown below), some minimal additions to the goat's head turned it into a caricature of the man in question. The body provides a background on which to plot the figures.

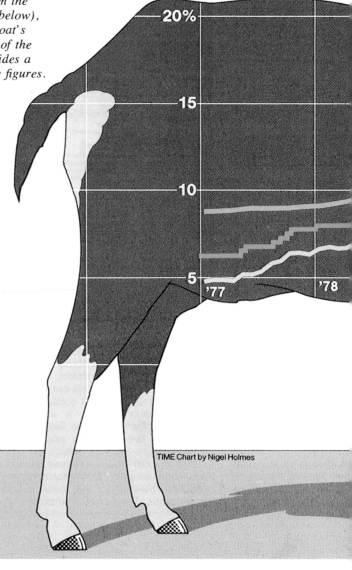

TIME Chart by Nigel Holmes

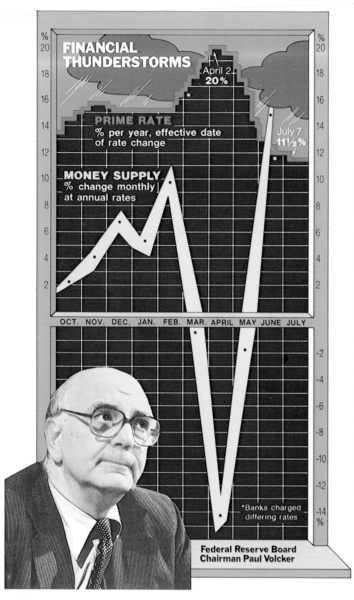

Federal Reserve Board
Chairman Paul Volcker

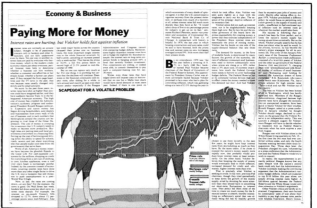

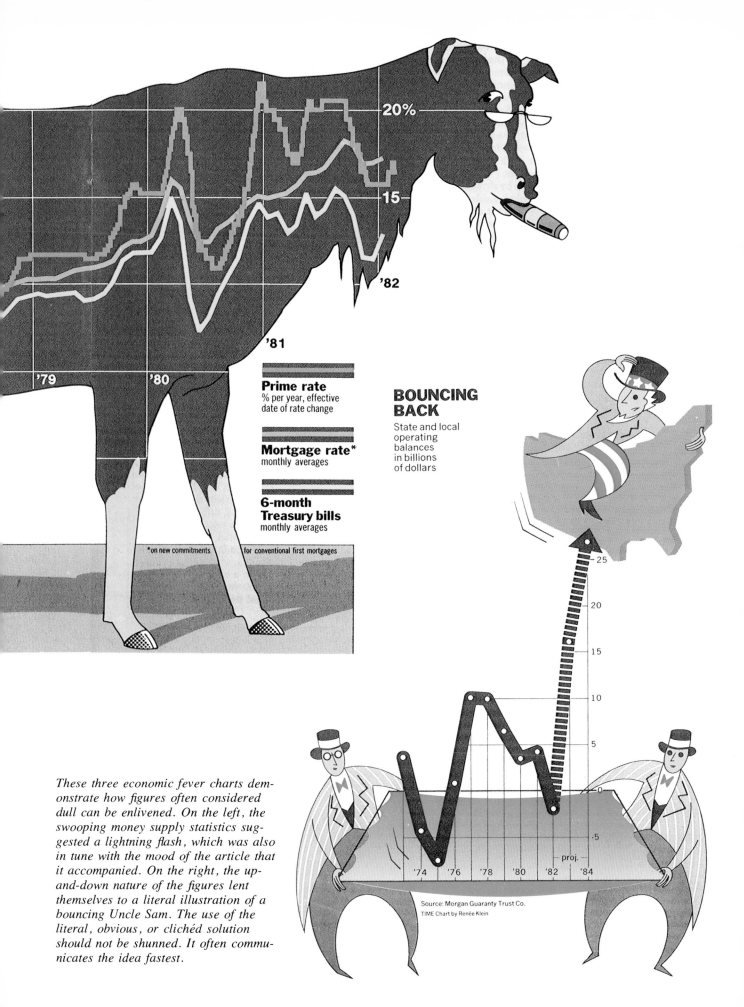

20%

15

'82

'81

'79   '80

Prime rate
% per year, effective
date of rate change

Mortgage rate*
monthly averages

**6-month
Treasury bills**
monthly averages

*on new commitments   for conventional first mortgages

**BOUNCING
BACK**

State and local
operating
balances
in billions
of dollars

25
20
15
10
5
0
.5

'74 '76 '78 '80 '82 '84
proj.

Source: Morgan Guaranty Trust Co.
TIME Chart by Renée Klein

These three economic fever charts demonstrate how figures often considered dull can be enlivened. On the left, the swooping money supply statistics suggested a lightning flash, which was also in tune with the mood of the article that it accompanied. On the right, the up-and-down nature of the figures lent themselves to a literal illustration of a bouncing Uncle Sam. The use of the literal, obvious, or clichéd solution should not be shunned. It often communicates the idea fastest.

THE AMERICAN NIGHTMARE
Median price of new one-family houses sold

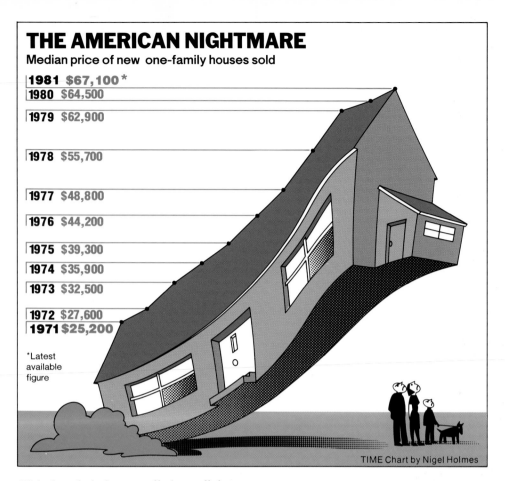

1981	$67,100*
1980	$64,500
1979	$62,900
1978	$55,700
1977	$48,800
1976	$44,200
1975	$39,300
1974	$35,900
1973	$32,500
1972	$27,600
1971	$25,200

*Latest available figure

TIME Chart by Nigel Holmes

With the whole house pulled up off the ground and looming over the family, a feeling about the difficulty of affording a home is conveyed. One pane of glass in each of the windows is darkened, forming subtle eyes that stare down at the people.

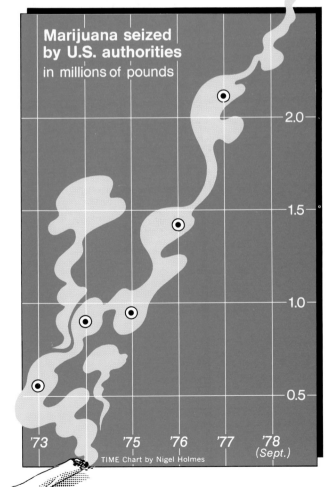

Marijuana seized by U.S. authorities
in millions of pounds

5.0
'78
(Sept.)

2.0

1.5

1.0

0.5

'73 '75 '76 '77 '78
(Sept.)

TIME Chart by Nigel Holmes

The smoke from the joint at the foot of this chart becomes a fever line. Breaking out of the rigid box and rising up like smoke itself, the sense of numbers growing fast is enhanced. The paleness of the smoke allowed it to be printed under the text of the magazine, thus integrating it into the page design.

PRIME INTEREST RATE
% per year, date of rate change

April 2
20%

Dec. 10
20%

July 25
11%

Nov. 20
1000.17

DOW JONES
industrials, daily closings

Oct. 31
917.75

Dec. 12
917.15

TIME Chart by Nigel Holmes

Two uncomfortable situations are illustrated here: (above) being caught between rising interest rates and falling stock prices and (right) a mailbox so high it has become impossible to post a letter. No grid is used in these two instances, but the figures and their dates are all printed so that there is no doubt about the information.

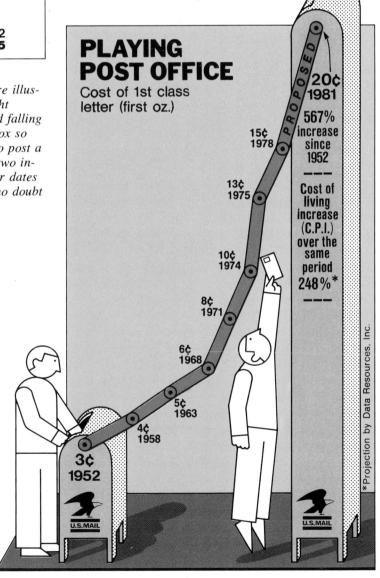

PLAYING POST OFFICE
Cost of 1st class letter (first oz.)

20¢ 1981 *PROPOSED*

567% increase since 1952

———

Cost of living increase (C.P.I.) over the same period 248%*

15¢ 1978

13¢ 1975

10¢ 1974

8¢ 1971

6¢ 1968

5¢ 1963

4¢ 1958

3¢ 1952

U.S. MAIL

U.S. MAIL

*Projection by Data Resources, Inc.

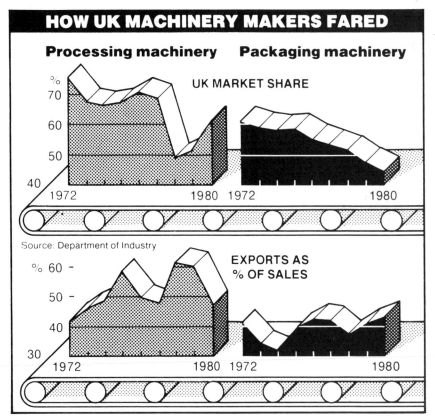

HOW UK MACHINERY MAKERS FARED

Processing machinery   Packaging machinery

UK MARKET SHARE

%
70
60
50
40
1972   1980 1972   1980

Source: Department of Industry

EXPORTS AS
% OF SALES

% 60 -
50 -
40
30
1972   1980 1972   1980

Well-labeled, simple three-dimensional fevers are arranged on conveyor belts. The shading on the front planes of the charts leaves no doubt about the reading points, and the projection into three dimensions therefore does not confuse, but underlines the idea of them as solid objects.

Elegance and precision mark this stark example. The very careful selection of line thickness gives the chart scientific accuracy and authority, and the overall design of the information into the square space is exciting while remaining extremely controlled.

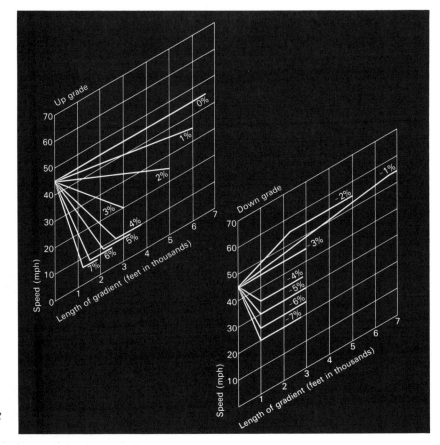

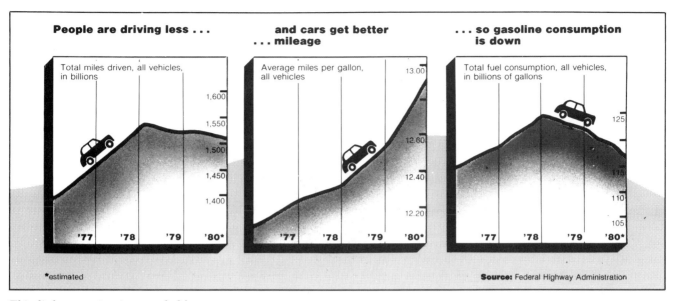

Source: Federal Highway Administration

This little story in pictures, held together by good headlining, is an example of excellent cooperation between the written and the drawn image.

The unaffected but well-designed presentation of business statistics is not a feature of many daily newspapers. This example from the financial pages shows that it can be done well. The use of a diagonal line screen in the background expertly shows the six charts in their best light.

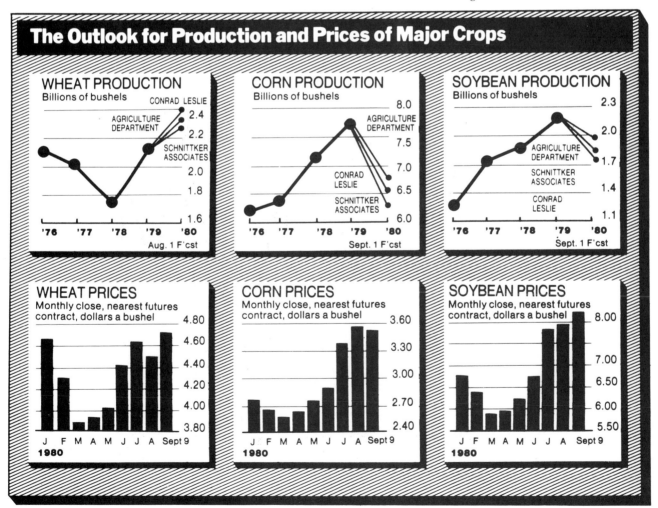

For a fever chart about education, what could be more natural than a chalkboard? A good merger between form and content, illustration and information.

A line-up of currencies is shown as fever charts featuring their value against the dollar. Their worried look confirms the information they carry towards the bathroom scales.

Value of each currency relative to the dollar; cumulative changes from December 1980.

Source: Federal Reserve Board

METROPOLITAN POLICE STRENGTH

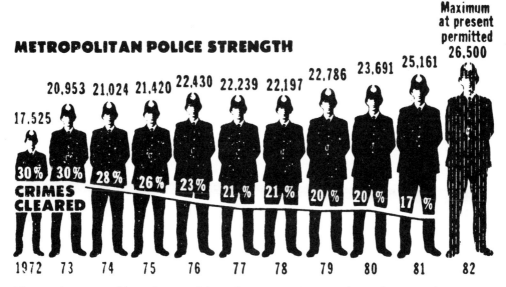

17,525 20,953 21,024 21,420 22,430 22,239 22,197 22,786 23,691 25,161

Maximum at present permitted 26,500

30% 30% 28% 26% 23% 21% 21% 20% 20% 17%

CRIMES CLEARED

1972 73 74 75 76 77 78 79 80 81 82

The combination of bar chart and fever line sets falling numbers of crimes cleared against the rising size of the police force. The visual interest generated and the amount of information packed into this small newspaper graphic are considerable. The handling of the type is excellent, especially the neat positioning of the percent sign as the fever line falls across the policemen's legs!

While on the subject, here's a horrifying chart about crime in America. The transformation of the map into a mask with rather threatening eye holes sets the scene for the dreadful rise in the numbers.

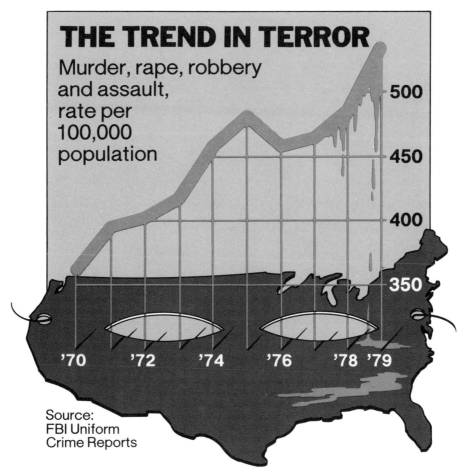

THE TREND IN TERROR

Murder, rape, robbery and assault, rate per 100,000 population

500

450

400

350

'70 '72 '74 '76 '78 '79

Source:
FBI Uniform
Crime Reports

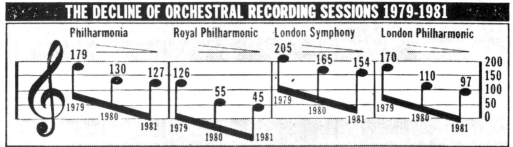

Gordon Beckett

Annual figures for recording and rehearsal sessions – in each case the dates are April to April

The "QUEEN MARY," MEASURED ALONG THE WATERLINE IS THE WORLD'S LONGEST SHIP

(Above left) On a number of counts, including its small size, great idea, and clarity of information, this newspaper bar chart ranks as one of my favorite examples in this book. Form and content come together perfectly. You instantly know what it is about, and the clear labeling takes you through the information easily.

(Left) From a children's textbook about the United States, this bar chart uses illustrations as the bars. While it can be criticized for enlarging or reducing the images in two dimensions instead of one (see page 174, in Chapter 6, Uses and Misuses), its young readers are not in fact going to be misled. The presentation adds so much interest—and in the case of the machine column, information—to the data that this far outweighs considerations of correctness of graphic methods.

Here is an example of a wonderful joining of illustration and statistics. By grouping together the world's tallest buildings of the time and then upending the Queen Mary and showing it alongside them, this artist has achieved a totally memorable image. This sort of visual comparison was much more popular in the 1930s than it is now; it should be used more.

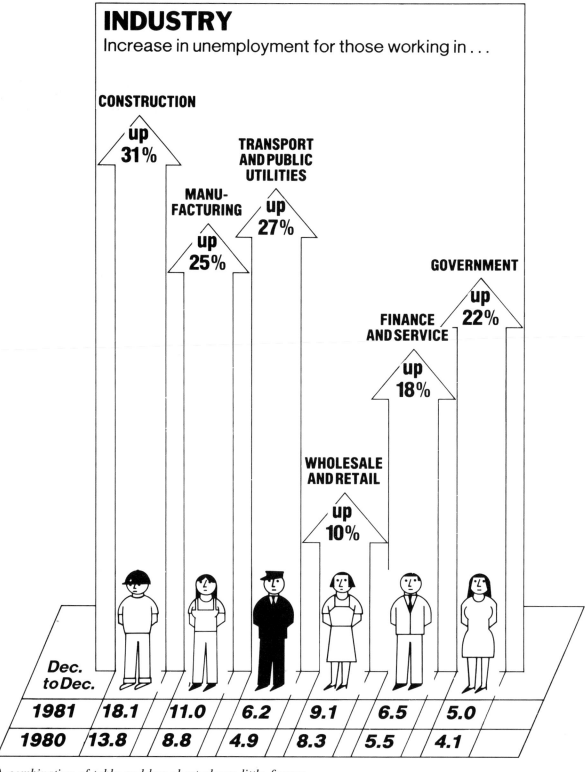

INDUSTRY
Increase in unemployment for those working in . . .

CONSTRUCTION
up 31%

MANU-FACTURING
up 25%

TRANSPORT AND PUBLIC UTILITIES
up 27%

GOVERNMENT
up 22%

FINANCE AND SERVICE
up 18%

WHOLESALE AND RETAIL
up 10%

Dec. to Dec.						
1981	18.1	11.0	6.2	9.1	6.5	5.0
1980	13.8	8.8	4.9	8.3	5.5	4.1

A combination of table and bar chart places little figures inside the bars to illustrate and humanize the information.

This population pyramid records the ages of Japanese people for the years 1935 and 1980. Totally without any decoration, it still holds the reader's interest and can be usefully studied for a long time.

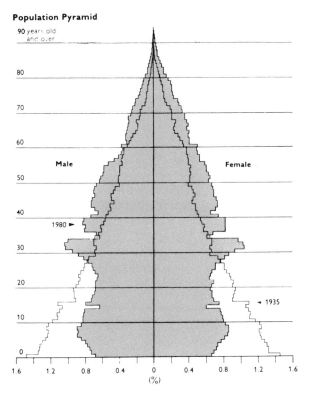

Population Pyramid

90 years old and over

80

70

60

Male   Female

50

40

1980 ►

30

20   ◄ 1935

10

0

1.6   1.2   0.8   0.4   0   0.4   0.8   1.2   1.6
(%)

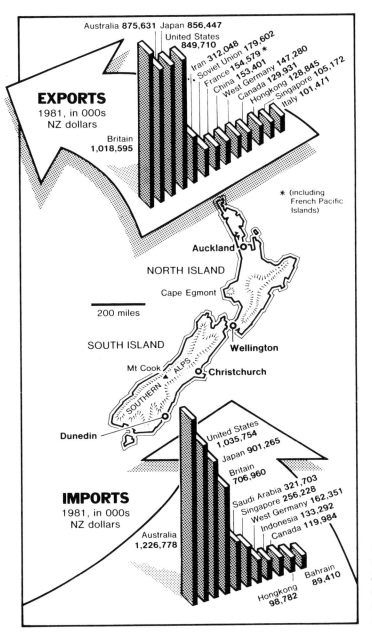

Australia **875,631** Japan **856,447**
United States **849,710**
Iran **312,048** Soviet Union **179,602**
France **154,579** *
China **153,401**
West Germany **147,280**
Canada **129,931**
Hongkong **128,845**
Singapore **105,172**
Italy **101,471**

EXPORTS
1981, in 000s
NZ dollars

Britain **1,018,595**

* (including French Pacific Islands)

Auckland

NORTH ISLAND

Cape Egmont

200 miles

SOUTH ISLAND

Mt Cook SOUTHERN ALPS **Wellington**
Christchurch

Dunedin

United States **1,035,754**
Japan **901,265**
Britain **706,960**
Saudi Arabia **321,703**
Singapore **256,228**
West Germany **162,351**
Indonesia **133,292**
Canada **119,984**

IMPORTS
1981, in 000s
NZ dollars

Australia **1,226,778**

Hongkong **98,782**

Bahrain **89,410**

What comes in and what goes out is clearly shown in these bar charts of imports and exports to and from New Zealand. With the arrangement of the arrows to the north and south of a map, a lot of material is displayed in a very economical space.

133

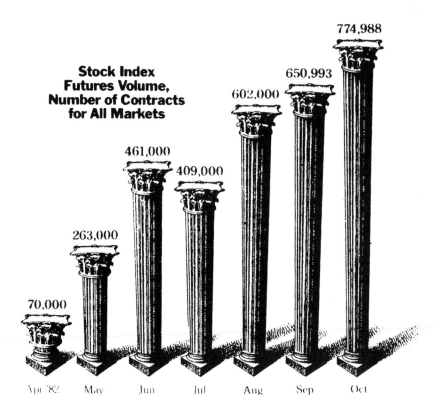

Stock Index Futures Volume, Number of Contracts for All Markets

70,000   263,000   461,000   409,000   602,000   650,993   774,988

Apr '82   May   Jun   Jul   Aug   Sep   Oct

These two instances feature illustrations as bars. In the column bar chart (above), the nice use of the solid, reliable bank/finance symbol is from a newspaper advertisement. (Below) spaghetti is stretched easily by the chartmaker. This advertisement transforms parallel lines into food and figures very elegantly.

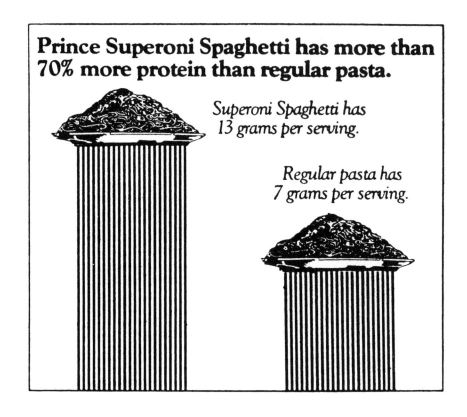

Prince Superoni Spaghetti has more than 70% more protein than regular pasta.

Superoni Spaghetti has 13 grams per serving.

Regular pasta has 7 grams per serving.

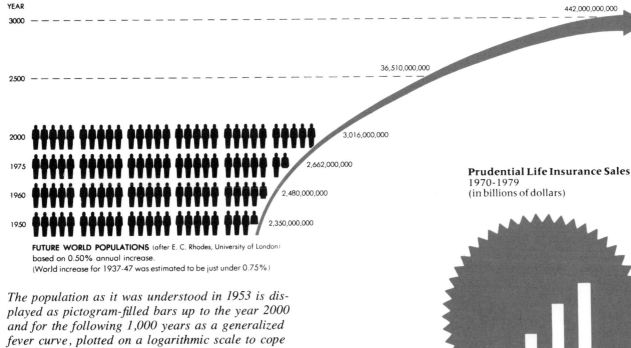

FUTURE WORLD POPULATIONS (after E. C. Rhodes, University of London) based on 0.50% annual increase.
(World increase for 1937-47 was estimated to be just under 0.75%)

The population as it was understood in 1953 is displayed as pictogram-filled bars up to the year 2000 and for the following 1,000 years as a generalized fever curve, plotted on a logarithmic scale to cope with the leap from 2 billion in 1950 to 442 billion in the year 3000. The ease with which a computer can generate repeated symbols will lead chart designers to make more use of multiple pictograms in the future.

Prudential Life Insurance Sales
1970-1979
(in billions of dollars)

1970   1973   1976   1979

The absolute simplicity of this bar chart from an annual report proves that impact can be made in a tiny space and with great economy of means.

By meticulously drawing a number of different houses and business premises and then stringing them together out of sequence, the artist here has created intriguing streets, which are then cut to length to show the frequency of burglaries in five counties.

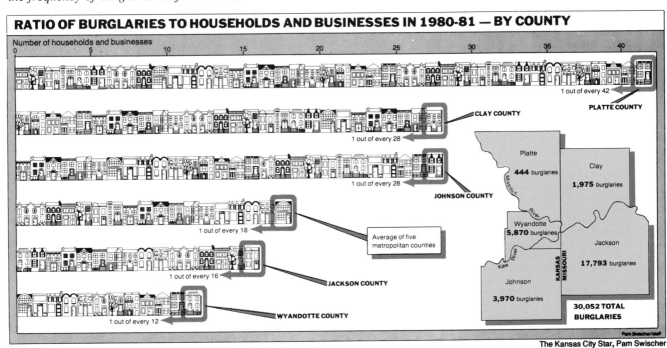

The Kansas City Star, Pam Swischer

For a chart about the buying of women's diamond jewelry (WDJ), a female hand is used as the base for a series of three-dimensional bars. From a large-scale presentation by an advertising agency to its diamond-manufacturing client, this chart indicates to which sector of the population their advertising should be directed.

W.D.J. Acquisition by Family Income

Female Family Heads, 1980

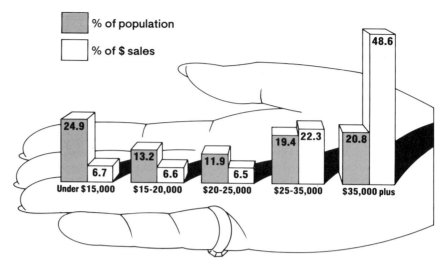

- ▨ % of population
- ☐ % of $ sales

	Under $15,000	$15-20,000	$20-25,000	$25-35,000	$35,000 plus
% of population	24.9	13.2	11.9	19.4	20.8
% of $ sales	6.7	6.6	6.5	22.3	48.6

6.9% of population, representing 12.2% of dollar sales, did not disclose income. Likely to be from upper bracket.

Good, symbolic line illustrations of farm buildings and cows (spot the differences!) present the information attractively in this newspaper graphic.

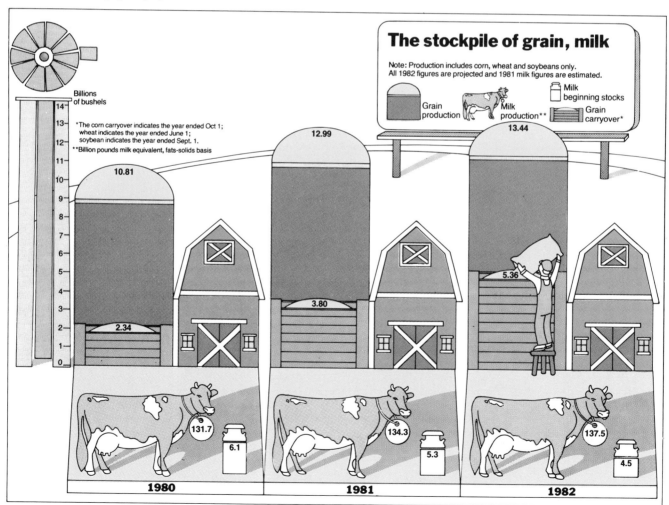

The stockpile of grain, milk

Note: Production includes corn, wheat and soybeans only.
All 1982 figures are projected and 1981 milk figures are estimated.

- Grain production
- Milk production**
- Milk beginning stocks
- Grain carryover*

Billions of bushels

*The corn carryover indicates the year ended Oct 1;
wheat indicates the year ended June 1;
soybean indicates the year ended Sept. 1.

**Billion pounds milk equivalent, fats-solids basis

	1980	1981	1982
Grain production	10.81	12.99	13.44
Grain carryover	2.34	3.80	5.36
Milk production	131.7	134.3	137.5
Milk beginning stocks	6.1	5.3	4.5

The Kansas City Times, Tom Dolphens

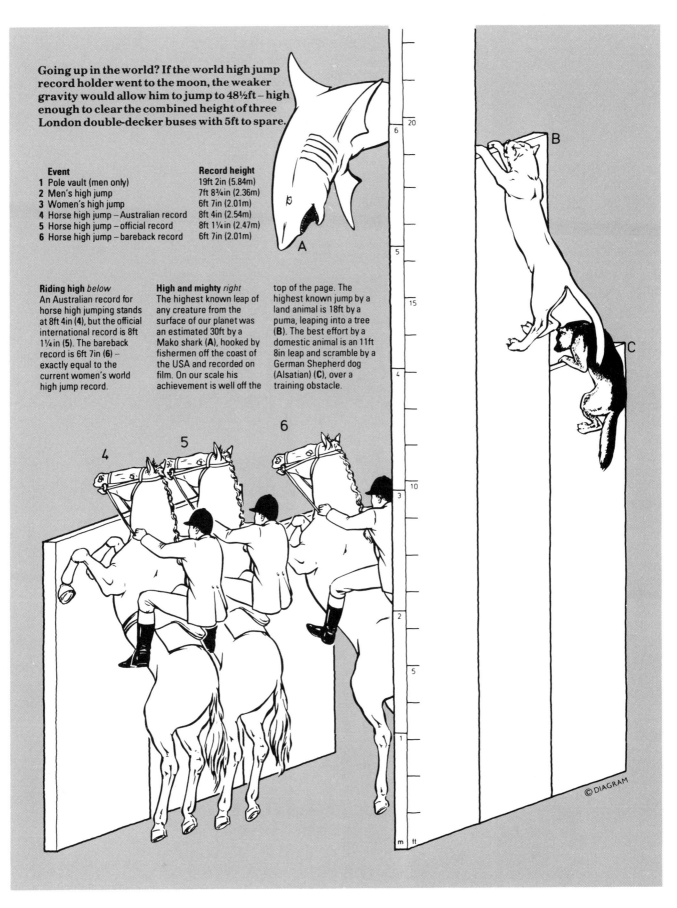

Going up in the world? If the world high jump record holder went to the moon, the weaker gravity would allow him to jump to 48½ft – high enough to clear the combined height of three London double-decker buses with 5ft to spare.

Event	Record height
1 Pole vault (men only)	19ft 2in (5.84m)
2 Men's high jump	7ft 8¾in (2.36m)
3 Women's high jump	6ft 7in (2.01m)
4 Horse high jump – Australian record	8ft 4in (2.54m)
5 Horse high jump – official record	8ft 1¼in (2.47m)
6 Horse high jump – bareback record	6ft 7in (2.01m)

Riding high *below*
An Australian record for horse high jumping stands at 8ft 4in (**4**), but the official international record is 8ft 1¼in (**5**). The bareback record is 6ft 7in (**6**) – exactly equal to the current women's world high jump record.

High and mighty *right*
The highest known leap of any creature from the surface of our planet was an estimated 30ft by a Mako shark (**A**), hooked by fishermen off the coast of the USA and recorded on film. On our scale his achievement is well off the top of the page. The highest known jump by a land animal is 18ft by a puma, leaping into a tree (**B**). The best effort by a domestic animal is an 11ft 8in leap and scramble by a German Shepherd dog (Alsatian) (**C**), over a training obstacle.

© DIAGRAM

From a book of comparisons, the bars here represent heights of objects reached by leaps from the ground (or ocean). An effective combination of fact-filled caption and straightforward line drawing results.

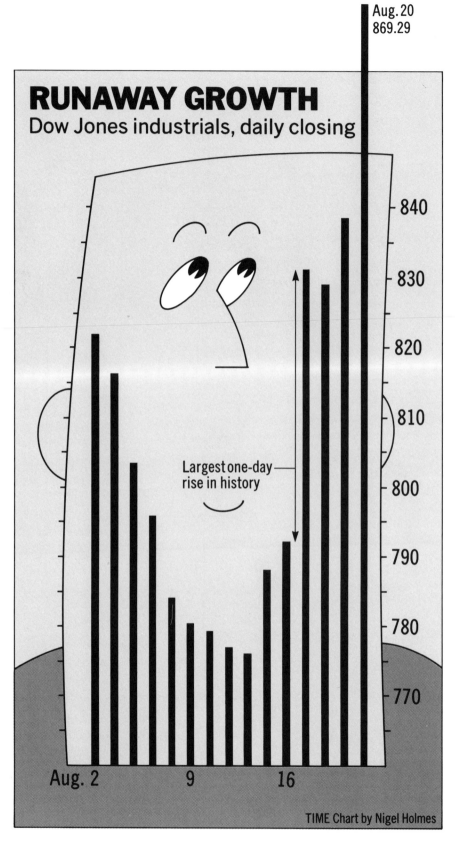

RUNAWAY GROWTH
Dow Jones industrials, daily closing

Aug. 20
869.29

840

830

820

810

800

790

780

770

Largest one-day
rise in history

Aug. 2   9   16

TIME Chart by Nigel Holmes

Making a face is a good way to humanize a set of statistics. A beard that grows upwards off the top of the head may be a little surreal, but it makes the point. Note how the placing of the words about one day's growth creates a moustache—as if he didn't have enough hair on his face already!

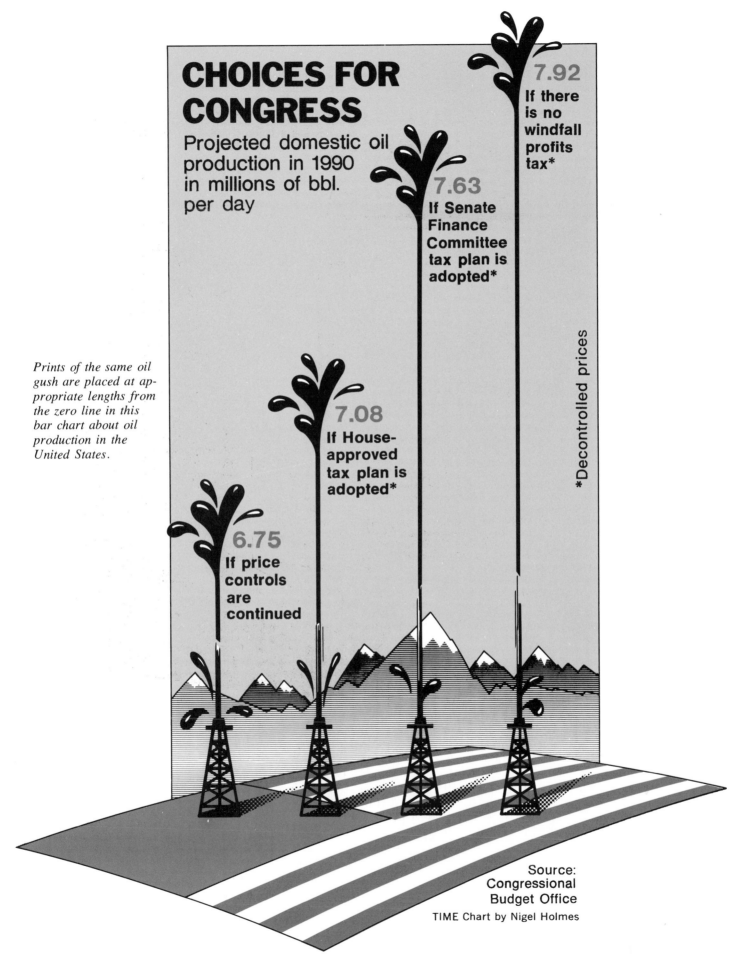

CHOICES FOR CONGRESS

Projected domestic oil production in 1990 in millions of bbl. per day

Prints of the same oil gush are placed at appropriate lengths from the zero line in this bar chart about oil production in the United States.

7.92
If there is no windfall profits tax*

7.63
If Senate Finance Committee tax plan is adopted*

7.08
If House-approved tax plan is adopted*

6.75
If price controls are continued

*Decontrolled prices

Source: Congressional Budget Office

TIME Chart by Nigel Holmes

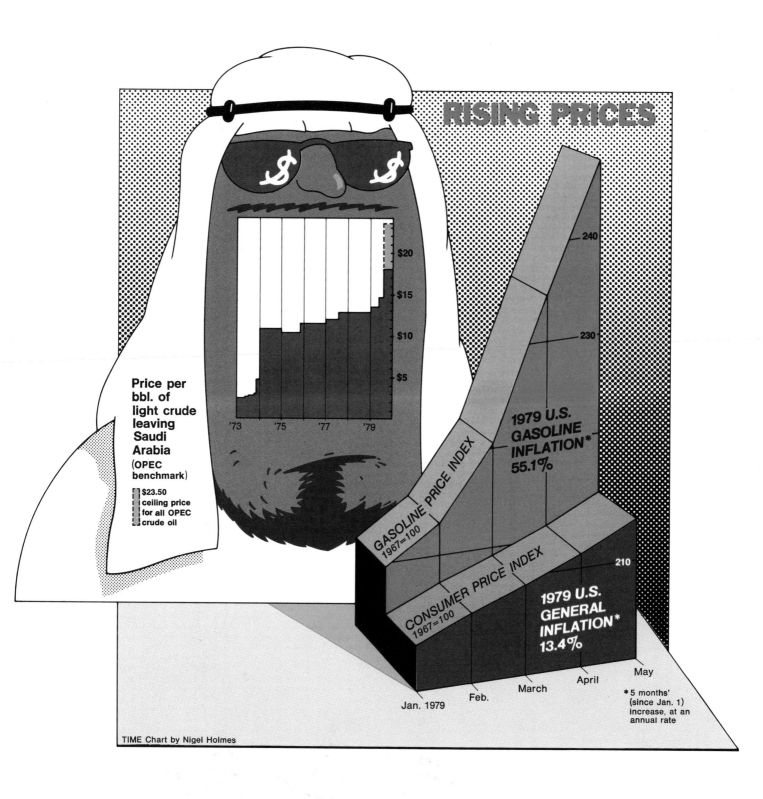

RISING PRICES

Price per
bbl. of
light crude
leaving
Saudi
Arabia
(OPEC
benchmark)

$23.50
ceiling price
for all OPEC
crude oil

$20
$15
$10
$5

'73   '75   '77   '79

GASOLINE PRICE INDEX
1967=100

1979 U.S.
GASOLINE
INFLATION*
55.1%

240

230

CONSUMER PRICE INDEX
1967=100

210

1979 U.S.
GENERAL
INFLATION*
13.4%

Jan. 1979   Feb.   March   April   May

*5 months'
(since Jan. 1)
increase, at an
annual rate

TIME Chart by Nigel Holmes

Oil—its price, production, consumption, and uses—is a subject that concerns us. OPEC's laughing Arab of 1979 (above) or rider galloping away with piles of money (right) gave way to a less happy situation for them in 1982 when they lost the lead in world oil production to non-OPEC countries (above right).

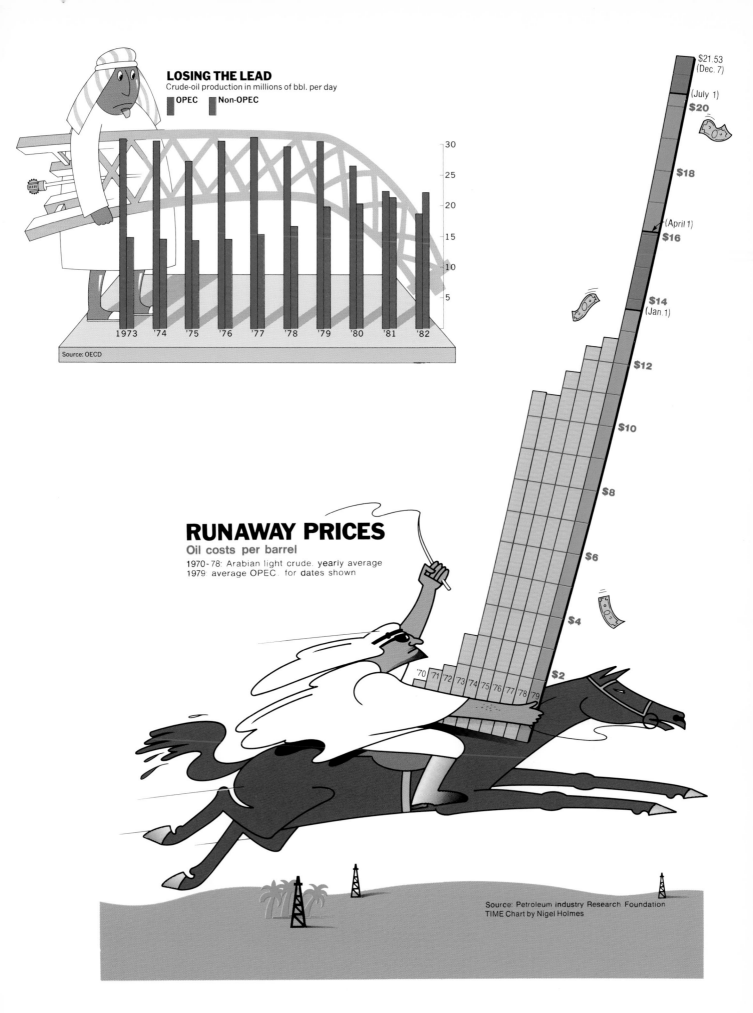

LOSING THE LEAD
Crude-oil production in millions of bbl. per day

■ OPEC ■ Non-OPEC

1973 '74 '75 '76 '77 '78 '79 '80 '81 '82

Source: OECD

RUNAWAY PRICES

Oil costs per barrel

1970-78: Arabian light crude, yearly average
1979: average OPEC, for dates shown

$21.53 (Dec. 7)
(July 1)
$20
$18
$16 (April 1)
$14 (Jan. 1)
$12
$10
$8
$6
$4
$2

'70 '71 '72 '73 '74 '75 '76 '77 '78 '79

Source: Petroleum Industry Research Foundation
TIME Chart by Nigel Holmes

141

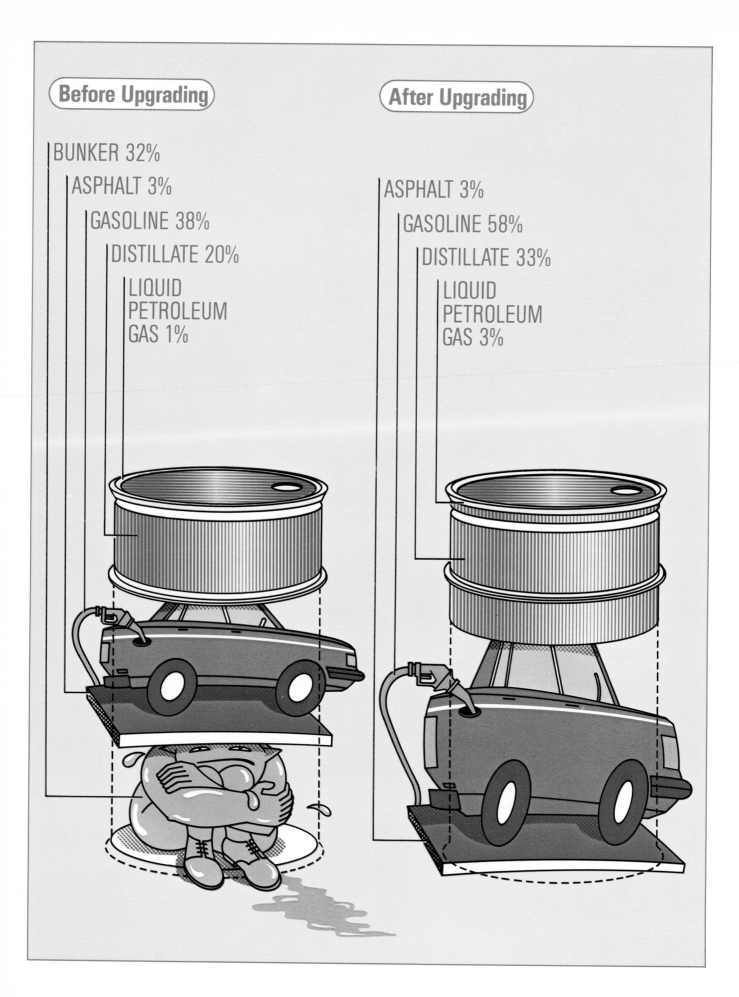

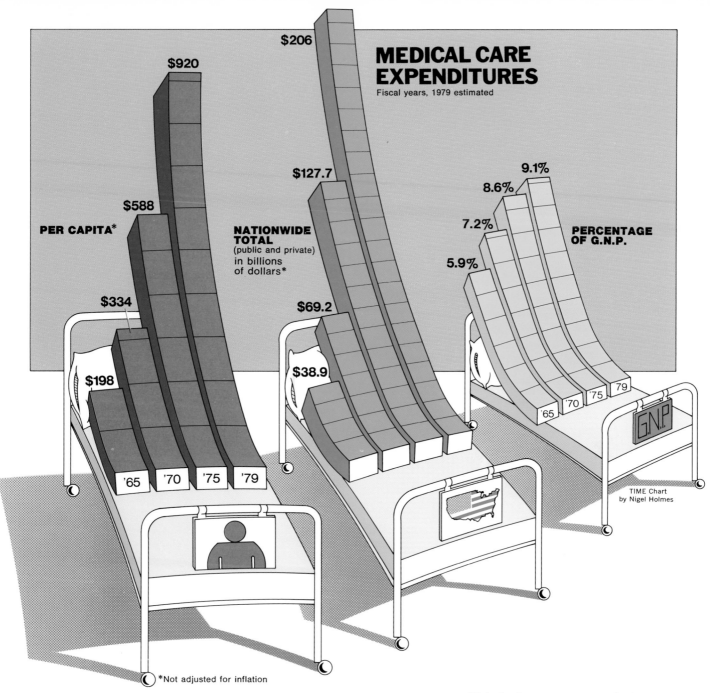

MEDICAL CARE EXPENDITURES
Fiscal years, 1979 estimated

PER CAPITA*

$920
$588
$334
$198

NATIONWIDE TOTAL
(public and private)
in billions
of dollars*

$206
$127.7
$69.2
$38.9

PERCENTAGE OF G.N.P.

9.1%
8.6%
7.2%
5.9%

'65 '70 '75 '79

G.N.P.

TIME Chart
by Nigel Holmes

*Not adjusted for inflation

With the bars sitting up in bed, an immediate visual impression of hospital costs is given. The exact figures are printed at the top of each bar so that there can be no confusion about the bends in them. Note the comments about beds on page 115.

This chart shows how oil is divided proportionately into different products before and after a process called "upgrading." The nasty little creature "bunker," or waste constituent at the bottom of the barrel, is eliminated after upgrading, which allows much more of the oil to be refined into useful products.

143

The 1981 Government Dollar (fiscal year 1981)

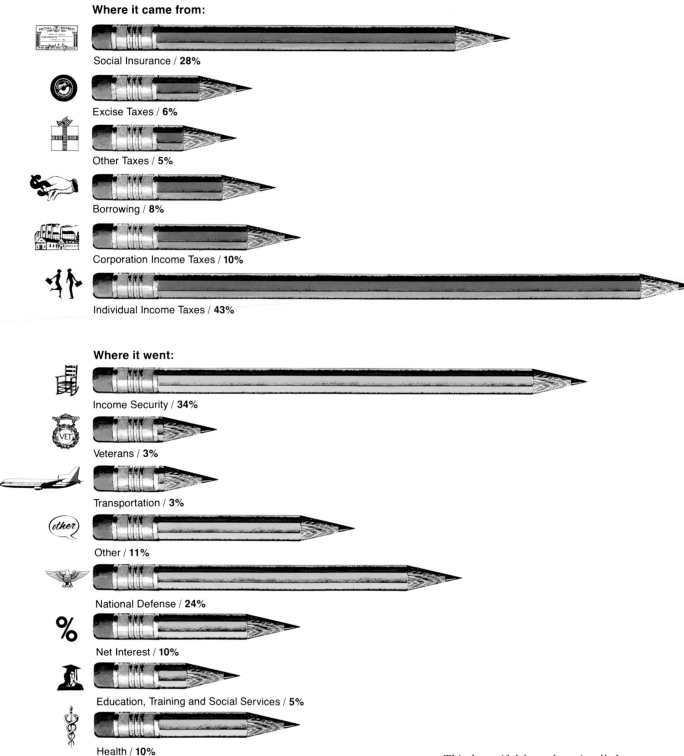

Where it came from:

Social Insurance / **28%**

Excise Taxes / **6%**

Other Taxes / **5%**

Borrowing / **8%**

Corporation Income Taxes / **10%**

Individual Income Taxes / **43%**

Where it went:

Income Security / **34%**

Veterans / **3%**

Transportation / **3%**

Other / **11%**

National Defense / **24%**

Net Interest / **10%**

Education, Training and Social Services / **5%**

Health / **10%**

This beautiful bar chart is all the more surprising in that it comes from an official government publication about taxes. It all goes to show that whatever they may be doing with our money which we may consider wrong, they are certainly charting it well! The color, the drawings, the restrained but perfectly clear typography, and the little string of symbols down the side all work well here.

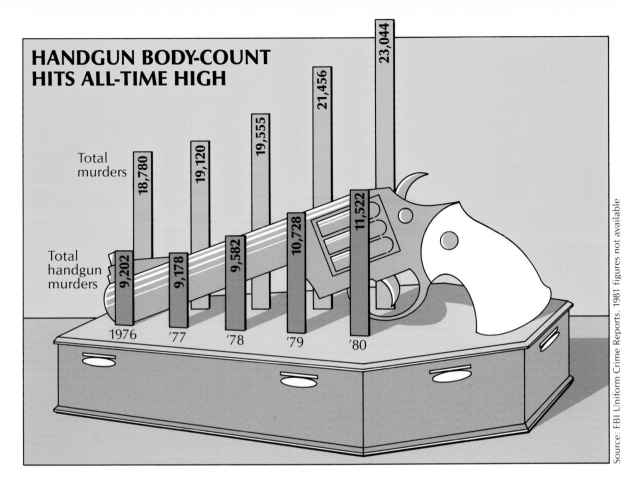

Source: FBI Uniform Crime Reports. 1981 figures not available

Two bar charts that use obvious but telling symbolism illustrate
the points made in a long magazine article about handgun control.

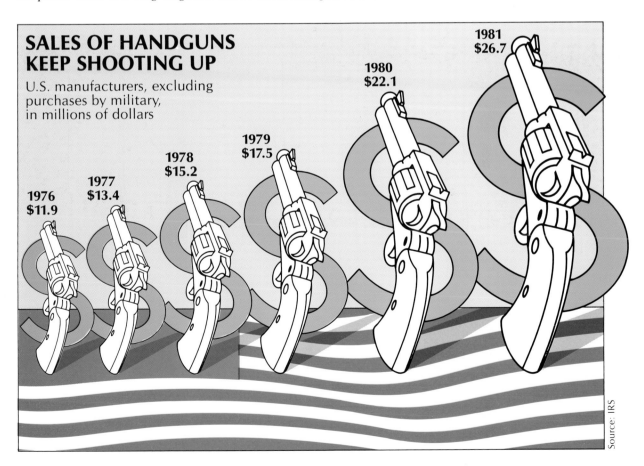

Source: IRS

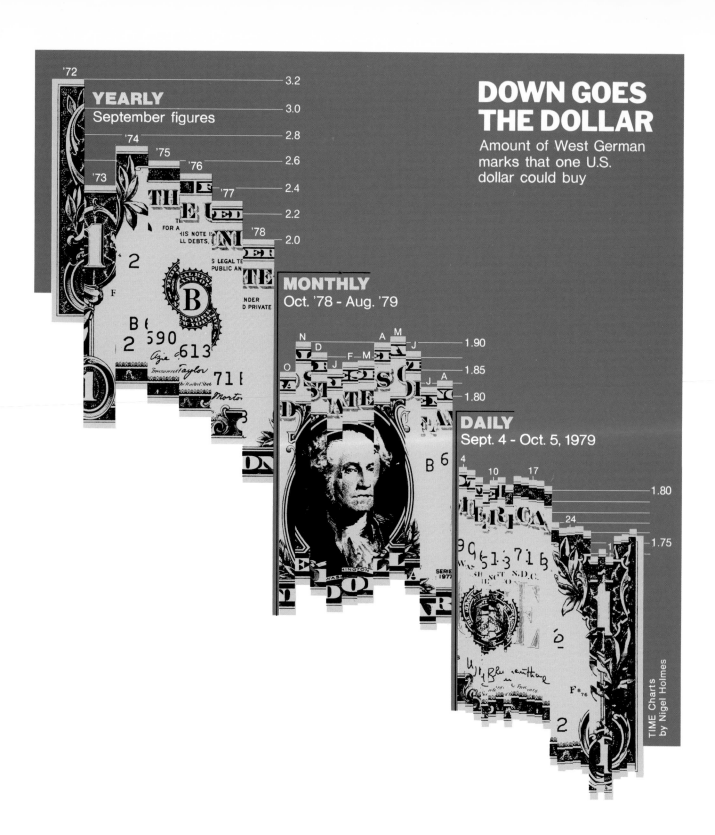

DOWN GOES THE DOLLAR
Amount of West German marks that one U.S. dollar could buy

YEARLY
September figures

MONTHLY
Oct. '78 - Aug. '79

DAILY
Sept. 4 - Oct. 5, 1979

TIME Charts
by Nigel Holmes

A dollar bill is cut up and the slices are used as bars to show, in three time stages, the dollar's slip against the West German mark.

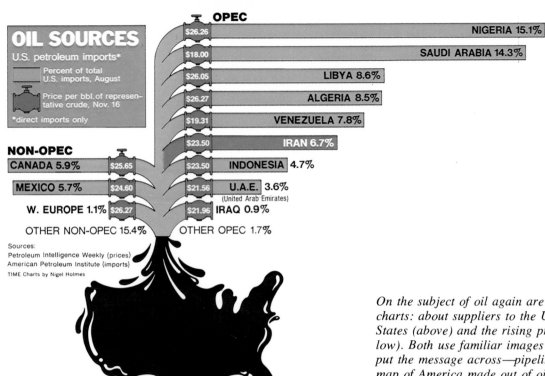

OIL SOURCES

U.S. petroleum imports*

◻ Percent of total U.S. imports, August

⬡ Price per bbl. of representative crude, Nov. 16

*direct imports only

OPEC

$26.26	NIGERIA 15.1%
$18.00	SAUDI ARABIA 14.3%
$26.05	LIBYA 8.6%
$26.27	ALGERIA 8.5%
$19.31	VENEZUELA 7.8%
$23.50	IRAN 6.7%
$23.50	INDONESIA 4.7%
$21.56	U.A.E. 3.6% (United Arab Emirates)
$21.96	IRAQ 0.9%

OTHER OPEC 1.7%

NON-OPEC

CANADA 5.9% $25.65

MEXICO 5.7% $24.60

W. EUROPE 1.1% $26.27

OTHER NON-OPEC 15.4%

Sources:
Petroleum Intelligence Weekly (prices)
American Petroleum Institute (imports)
TIME Charts by Nigel Holmes

On the subject of oil again are two bar charts: about suppliers to the United States (above) and the rising price (below). Both use familiar images to help put the message across—pipelines and a map of America made out of oil in one and a cash register/gasoline pump in the other.

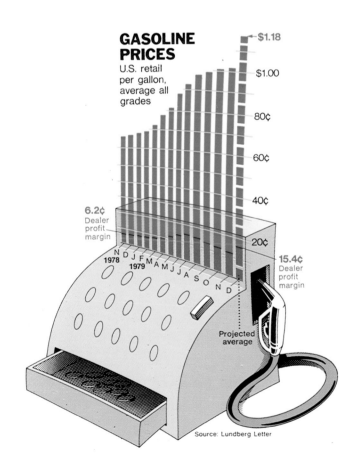

GASOLINE PRICES

U.S. retail per gallon, average all grades

$1.18
$1.00
80¢
60¢
40¢
20¢

6.2¢ Dealer profit margin

15.4¢ Dealer profit margin

N 1978 D J F M A M J J A S O N D 1979

Projected average

Source: Lundberg Letter

147

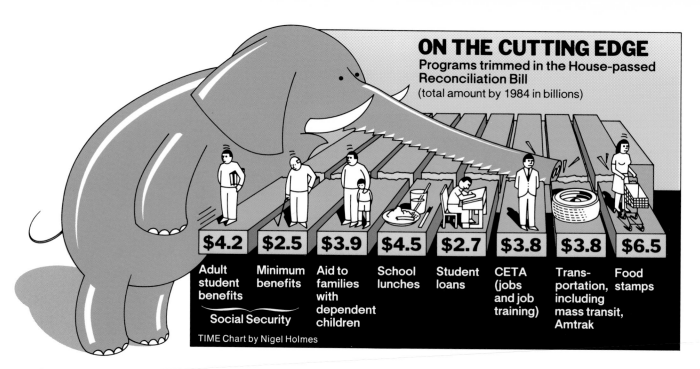

ON THE CUTTING EDGE
Programs trimmed in the House-passed
Reconciliation Bill
(total amount by 1984 in billions)

$4.2	$2.5	$3.9	$4.5	$2.7	$3.8	$3.8	$6.5
Adult student benefits	Minimum benefits	Aid to families with dependent children	School lunches	Student loans	CETA (jobs and job training)	Transportation, including mass transit, Amtrak	Food stamps
Social Security							

TIME Chart by Nigel Holmes

The Republican elephant saws billions of dollars from social programs whose recipients are precariously positioned on them. This "bar" chart is really more a table in that there is no charting of the data, merely written information about the amounts involved.

**Percentage of Recruiter Contacts
in Response to
Direct Mail Program**

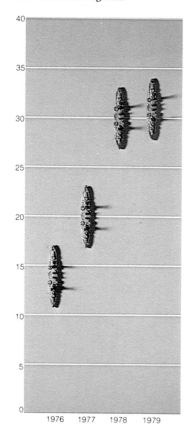

This photographic chart from an annual report uses toy ships as modular elements and is a telling use of a tiny space.

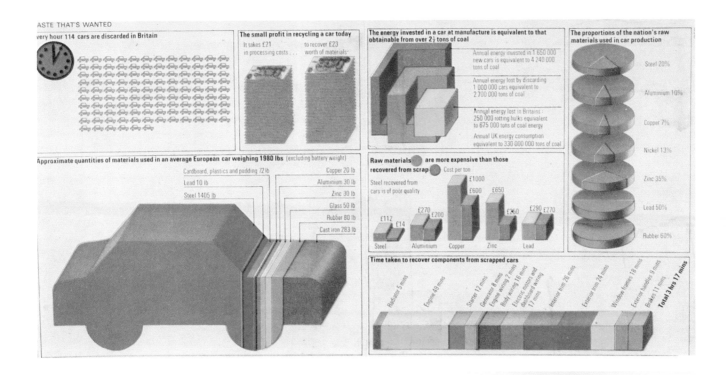

ASTE THAT'S WANTED

very hour 114 cars are discarded in Britain

The small profit in recycling a car today
It takes £21 in processing costs . . . to recover £23 worth of materials

The energy invested in a car at manufacture is equivalent to that obtainable from over 2½ tons of coal

Annual energy invested in 1 650 000 new cars is equivalent to 4 240 000 tons of coal

Annual energy lost by discarding 1 000 000 cars equivalent to 2 700 000 tons of coal

Annual energy lost in Britains 250 000 rotting hulks equivalent to 675 000 tons of coal energy

Annual UK energy consumption equivalent to 330 000 000 tons of coal

The proportions of the nation's raw materials used in car production

Steel 20%
Aluminium 10%
Copper 7%
Nickel 13%
Zinc 35%
Lead 50%
Rubber 60%

Approximate quantities of materials used in an average European car weighing 1980 lbs (excluding battery weight)

Cardboard, plastics and padding 72 lb
Lead 10 lb
Steel 1405 lb
Copper 20 lb
Aluminium 30 lb
Zinc 30 lb
Glass 50 lb
Rubber 80 lb
Cast iron 283 lb

Raw materials are more expensive than those recovered from scrap Cost per ton

Steel recovered from cars is of poor quality

	£1000			
£112 £14	£270 £200	£600	£650 £260	£290 £270
Steel	Aluminium	Copper	Zinc	Lead

Time taken to recover components from scrapped cars

Radiator 5 mins · Engine 49 mins · Starter 17 mins · Generator 8 mins · Engine wiring 2 mins · Body wiring 16 mins · Electric motors and dashboard wiring 17 mins · Interior trim 28 mins · Exterior trim 24 mins · Window frames 18 mins · Exterior handles 9 mins · Brakes 11 mins · **Total 3 hrs 17 mins**

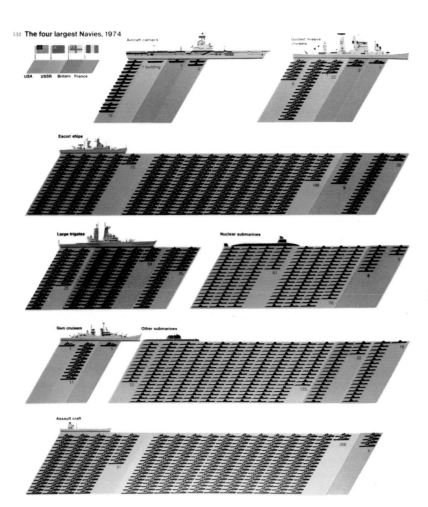

132 **The four largest Navies, 1974**

USA USSR Britain France

Aircraft carriers · 1 building
Guided missile cruisers
Escort ships
Large frigates
Nuclear submarines
Gun cruisers
Other submarines
Assault craft

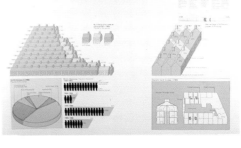

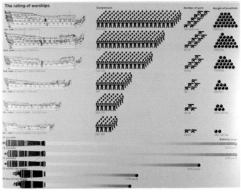

The rating of warships

A group of well-designed pages use all kinds of graphic presentation, both illustrated and abstract.

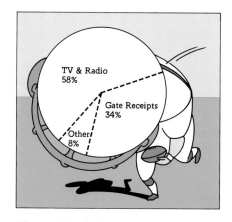 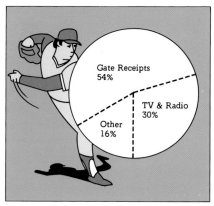 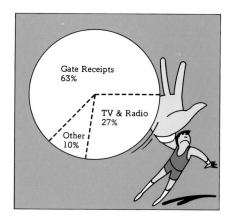

Three related charts are about money and sports. An element of the particular sport is used as the divided circle to show where the huge amounts come from for each one.

A pie about the cost of making cosmetics is divided firstly into the retailer's and the manufacturer's share and then subdivided within the latter. By removing the 8-cent share of the product itself from the pie and positioning it where it is used, a nice, illustrated element is introduced.

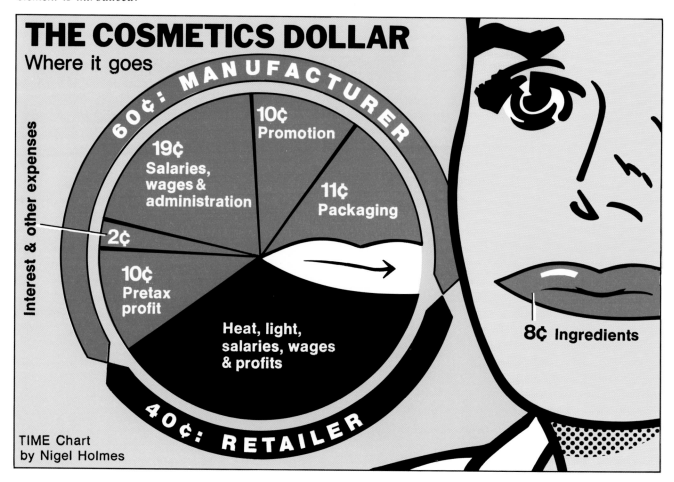

To show who owns the world's gold, two stylized hands are drawn literally holding their slices of the pie, while a fairly sizable chunk that is either lost or undetermined floats between the two.

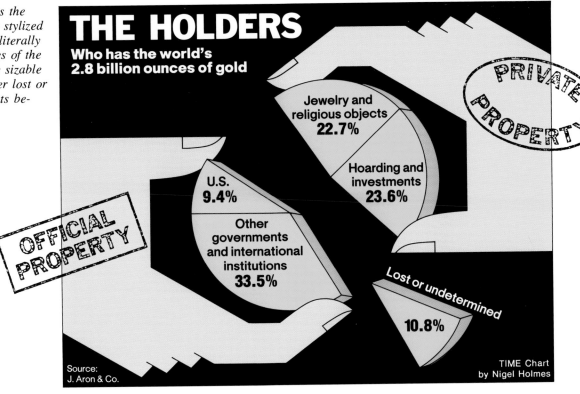

THE HOLDERS
Who has the world's 2.8 billion ounces of gold

PRIVATE PROPERTY

Jewelry and religious objects
22.7%

Hoarding and investments
23.6%

U.S.
9.4%

OFFICIAL PROPERTY

Other governments and international institutions
33.5%

Lost or undetermined
10.8%

Source: J. Aron & Co.

TIME Chart by Nigel Holmes

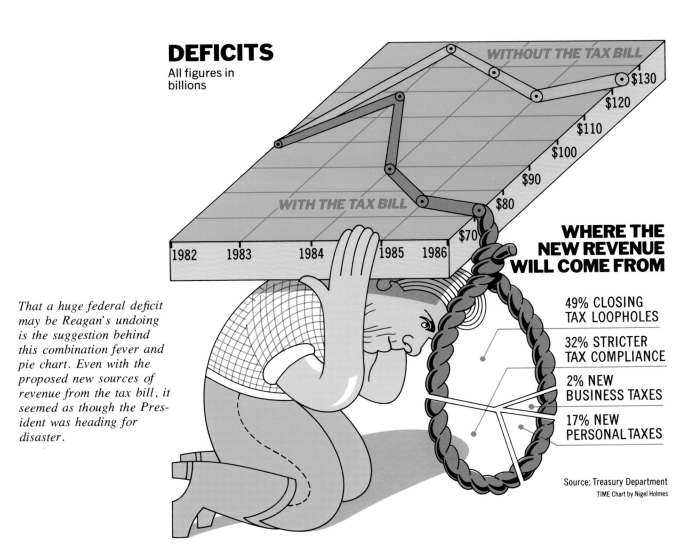

DEFICITS
All figures in billions

WITHOUT THE TAX BILL

$130
$120
$110
$100
$90
$80
$70

WITH THE TAX BILL

1982 1983 1984 1985 1986

WHERE THE NEW REVENUE WILL COME FROM

49% CLOSING TAX LOOPHOLES

32% STRICTER TAX COMPLIANCE

2% NEW BUSINESS TAXES

17% NEW PERSONAL TAXES

That a huge federal deficit may be Reagan's undoing is the suggestion behind this combination fever and pie chart. Even with the proposed new sources of revenue from the tax bill, it seemed as though the President was heading for disaster.

Source: Treasury Department
TIME Chart by Nigel Holmes

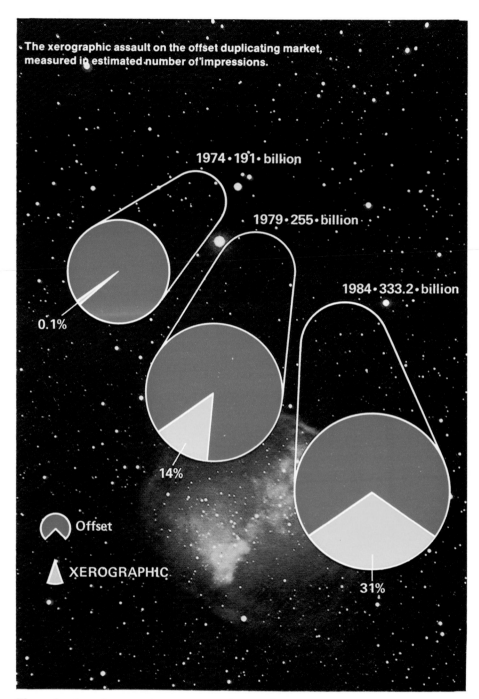

*These floating abstract pies are set against a photograph
of distant galaxies, conveying an impression of the future.*

How Many Are Idled by Industrial Decline?

8 million U.S. workers had lost their jobs as of January 1983

Of whom 1.6 million were in declining industries

Of whom 240,000 were out of work 26 weeks or more

Of whom 60,000 had 10 years or more on the job

Though not a pie chart by most definitions, this well-designed diagram shows layers of information about 8 million unemployed United States workers. They are graphically positioned in a hole to represent their plight.

1983

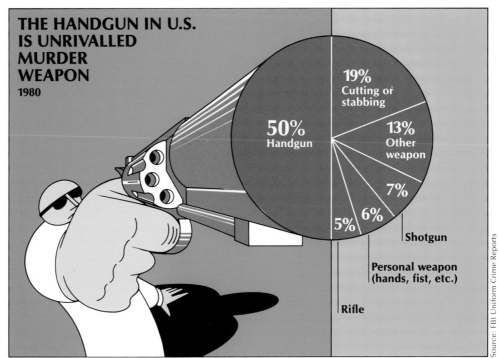

THE HANDGUN IN U.S. IS UNRIVALLED MURDER WEAPON
1980

50% Handgun

19% Cutting or stabbing

13% Other weapon

7%

6%

5%

Shotgun

Personal weapon (hands, fist, etc.)

Rifle

From the same series as the two bar charts on page 145, this pie demonstrates how many more murders are committed with handguns than with any other weapon. The dramatic angle of the gun pointing directly at the reader brings the point home quickly.

Source: FBI Uniform Crime Reports

TELLTALES OF TWO CITIES

Cost of living

	New York City	Moscow
Hours worked per week	39.7	42
Manufacturing worker's earnings per week	$265.60	$56.54
Monthly rental 3-room apt.	$1,000	$37
Heat and electricity per month	$82	$4.50
Bus or subway per ride	50¢	8¢
Car	$6,200 (Citation)	$10,000 (Zhiguli)
Gallon of gasoline	$1.35	$1.25
Vodka (1 liter)	$6	$11
Dental checkup	$32	Free
Ballpoint pen	29¢	$1.50
1 lb. chicken	66¢	$2.55
Pack of cigarettes	75¢	52¢
Loaf of bread (1 lb.)	62¢	24¢
Jeans	$18.50	$45
Man's leather shoes	$42	$45
Vacation (2 weeks per person)*	$910	$120
Woman's dress	$90	$60
Sofa	$600	$260
Gold wedding ring	$75	$225
Hard-cover novel	$12.95	$3
Color television	$710 (25-in. screen)	$1,094 (24-in. screen)
Pantyhose	$1.50	$10
Newspaper	25¢ (Daily News)	5¢ (Izvestiya)
Woman's leather shoes	$33	$40

All figures are typical current prices.

* Vacations: the New York figure represents a peak season cost with meals at a major Florida resort city; 70% of the Moscow figure is paid by the worker's union.

TIME Chart by Nigel Holmes

1982 projections	GROWTH % change in real G.N.P., 4th Q. over 4th Q.	INFLATION % increase in C.P.I., Dec. over Dec.	UNEMPLOYMENT as % of civilian labor force, year-end
W. GERMANY	-2.5%	4.5%	7.8%
FRANCE	+0.2%	9.8%	8.9%
BRITAIN	0	6%	12.9%
ITALY	+1%	16.5%	9%
W. EUROPE	+0.5%	8.6%	10.5%

1983 FORECAST by TIME's European Board of Economists

	GROWTH		INFLATION		UNEMPLOYMENT	
W. GERMANY	+2%		3%		9.5%	
FRANCE	+1%		9%		9.8%	
BRITAIN	+2.3%		5.8%		14%	
ITALY*	-0.8%	+1.5%	14%	21%	11.5%	9%
W. EUROPE	+2.5%		7.5%		11%	

As an addition to this table of economic facts with projections of future performance, a map of the countries involved has been drawn and given human features: crutches to convey something of the sorry state of their economies at that time and worries heaped on their shoulders.

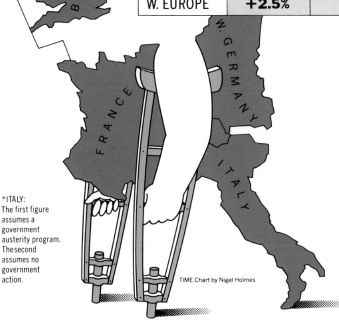

*ITALY:
The first figure assumes a government austerity program. The second assumes no government action.

TIME Chart by Nigel Holmes

Good, simple use of symbols enlivens this small black-and-white newspaper table about unemployment.

Jobless rate by industry

(Seasonally adjusted rate for February)

Mining and metals **18.4**

Construction **19.7**

Manufacturing **13.3**

Services and finance **7.3**

Government **6.0**

Wholesale and retail trade **10.9**

Transportation and public utilities **8.0**

AP/News Graphics

SOURCE: Labor Department

This table shows the difference in the cost of living in New York and Moscow in 1980. It is a real tribute to the researcher's art of finding and checking compatible facts and figures from two sides of the globe. Note, for instance, that chickens are the same the world over, but bread in Russia simply doesn't come in the shape familiar to Western nations.

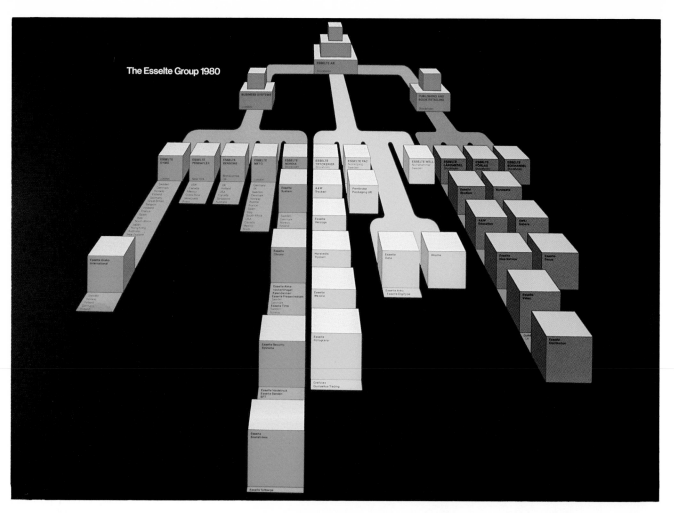

The Esselte Group 1980

Two organizational tables show the path of responsibilities within companies. They demonstrate how a highly organized approach to the design of tabular material can be an end in itself without the addition of any illustrative or unnecessary decorative elements.

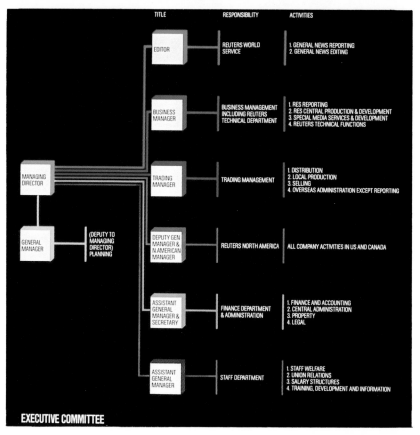

TITLE	RESPONSIBILITY	ACTIVITIES
EDITOR	REUTERS WORLD SERVICE	1. GENERAL NEWS REPORTING 2. GENERAL NEWS EDITING
BUSINESS MANAGER	BUSINESS MANAGEMENT INCLUDING REUTERS TECHNICAL DEPARTMENT	1. RES REPORTING 2. RES CENTRAL PRODUCTION & DEVELOPMENT 3. SPECIAL MEDIA SERVICES & DEVELOPMENT 4. REUTERS TECHNICAL FUNCTIONS
TRADING MANAGER	TRADING MANAGEMENT	1. DISTRIBUTION 2. LOCAL PRODUCTION 3. SELLING 4. OVERSEAS ADMINISTRATION EXCEPT REPORTING
DEPUTY GEN MANAGER & N AMERICAN MANAGER	REUTERS NORTH AMERICA	ALL COMPANY ACTIVITIES IN US AND CANADA
ASSISTANT GENERAL MANAGER & SECRETARY	FINANCE DEPARTMENT & ADMINISTRATION	1. FINANCE AND ACCOUNTING 2. CENTRAL ADMINISTRATION 3. PROPERTY 4. LEGAL
ASSISTANT GENERAL MANAGER	STAFF DEPARTMENT	1. STAFF WELFARE 2. UNION RELATIONS 3. SALARY STRUCTURES 4. TRAINING, DEVELOPMENT AND INFORMATION

MANAGING DIRECTOR

GENERAL MANAGER — (DEPUTY TO MANAGING DIRECTOR) PLANNING

EXECUTIVE COMMITTEE

CRISIS WITH THE SOVIETS

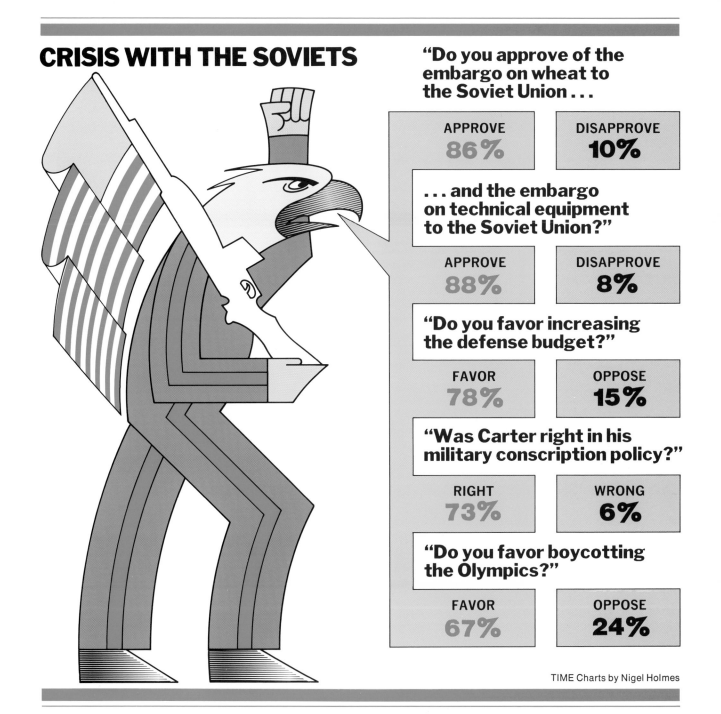

"Do you approve of the embargo on wheat to the Soviet Union . . .

APPROVE	DISAPPROVE
86%	10%

. . . and the embargo on technical equipment to the Soviet Union?"

APPROVE	DISAPPROVE
88%	8%

"Do you favor increasing the defense budget?"

FAVOR	OPPOSE
78%	15%

"Was Carter right in his military conscription policy?"

RIGHT	WRONG
73%	6%

"Do you favor boycotting the Olympics?"

FAVOR	OPPOSE
67%	24%

TIME Charts by Nigel Holmes

The results of a poll during 1980 indicated an overwhelmingly hawkish mood in the United States. The answers are therefore shown coming from an aggressively militant bird speaking into a stylized speech balloon.

Looking forward to the future of four European countries' economic growth is the theme of this simple table. The idea is literal and gets to the point quickly.

These nice arrows with their shadows to make them sit up on the page work very well for the left, right, or center politics of the various countries. It is interesting to see that of the 24 countries only eight have not changed political direction in the 10 years covered.

The melting away of assets that had been frozen during the hostage crisis of 1979–1980 formed the basis of this complicated table/flowchart. The gradation of color in the background is created by applying an adhesive-backed gradated screen to the artwork and printing it as red line over a yellow tint.

INFLATION
Change in C.P.I.
Dec. over Dec.

	1981	FORECAST 1982	
BRITAIN	12%	8%	
FRANCE	14%	10.5%	
ITALY	18.2%	16%	
W. GERMANY	6.3%	4%	
U.S.	8.9%	6.5%	

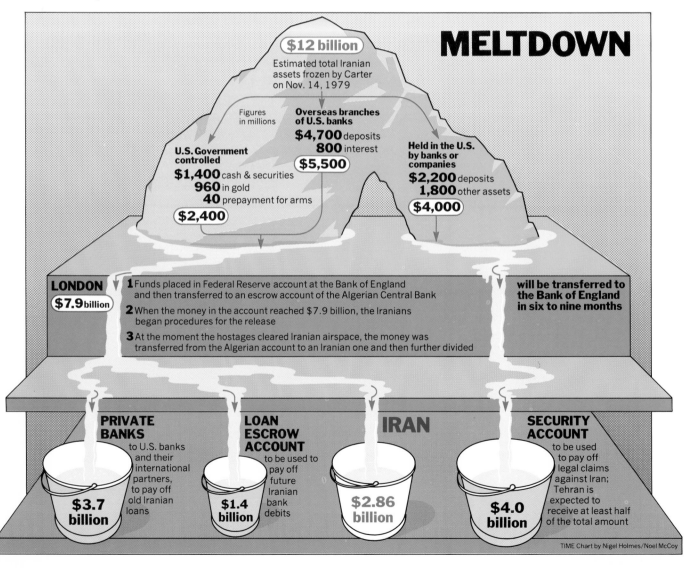

MELTDOWN

$12 billion
Estimated total Iranian assets frozen by Carter on Nov. 14, 1979

Figures in millions

Overseas branches of U.S. banks
$4,700 deposits
800 interest
$5,500

U.S. Government controlled
$1,400 cash & securities
960 in gold
40 prepayment for arms
$2,400

Held in the U.S. by banks or companies
$2,200 deposits
1,800 other assets
$4,000

LONDON
$7.9 billion

1 Funds placed in Federal Reserve account at the Bank of England and then transferred to an escrow account of the Algerian Central Bank

2 When the money in the account reached $7.9 billion, the Iranians began procedures for the release

3 At the moment the hostages cleared Iranian airspace, the money was transferred from the Algerian account to an Iranian one and then further divided

will be transferred to the Bank of England in six to nine months

PRIVATE BANKS to U.S. banks and their international partners, to pay off old Iranian loans
$3.7 billion

LOAN ESCROW ACCOUNT to be used to pay off future Iranian bank debits
$1.4 billion

IRAN
$2.86 billion

SECURITY ACCOUNT to be used to pay off legal claims against Iran; Tehran is expected to receive at least half of the total amount
$4.0 billion

TIME Chart by Nigel Holmes/Noel McCoy

Where Governments Stand

A Survey of the Political Direction of
Governments in 24 Countries

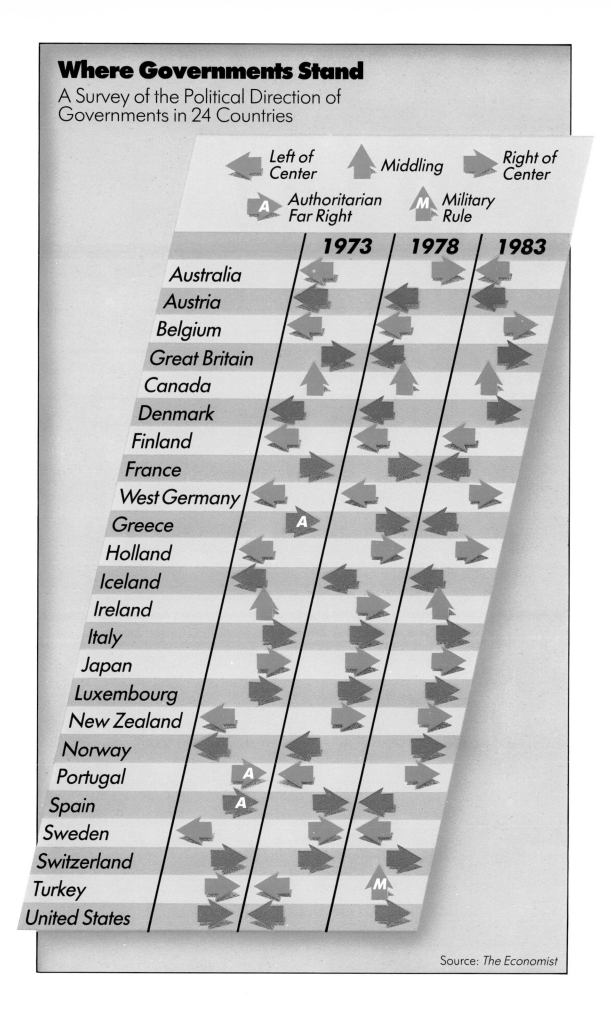

Source: *The Economist*

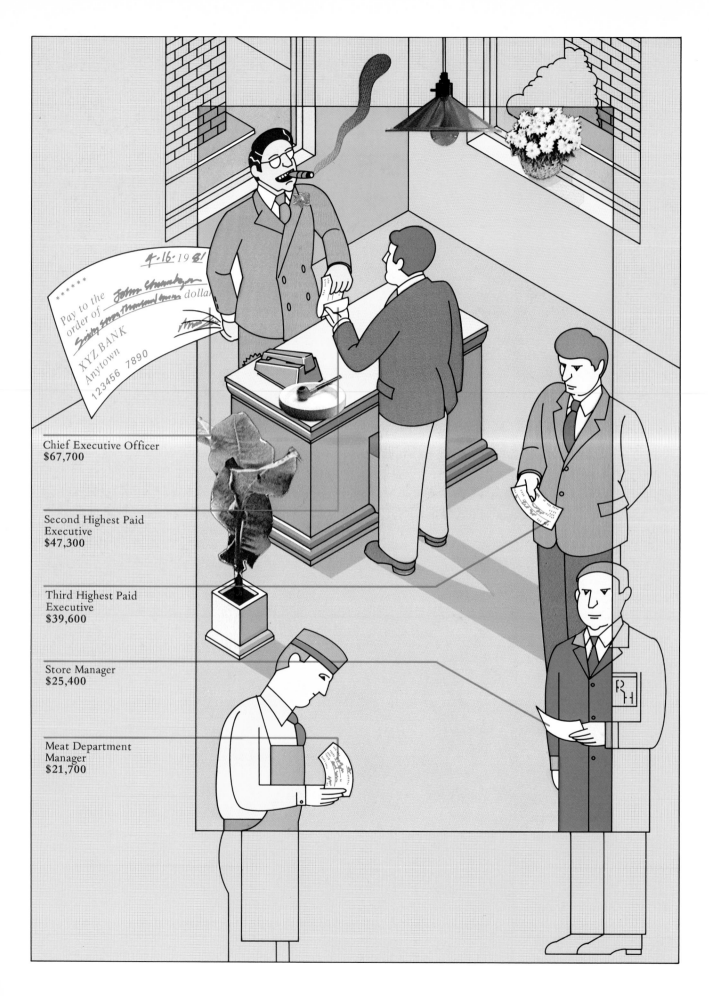

Chief Executive Officer
$67,700

Second Highest Paid
Executive
$47,300

Third Highest Paid
Executive
$39,600

Store Manager
$25,400

Meat Department
Manager
$21,700

(Left) Where there are only a few facts to be included in a full-page table a decision may be made to make more of the illustration than one normally would. Here the outside part of the page contains the information and the inside, with its collaged photographs, is an illustration of the hierarchy in a food store chain.

(Right) This newspaper table of football playoffs of necessity had to be designed before the information to go into it was available. As such it provides a most effective framework into which the names of the teams could be fitted. The bold use of the third dimension on the boxes is particularly good.

(Below) This excellent collaboration between illustrator and cartographer/ chartmaker is included as a table by virtue of the mass of information laid out on it. The very clever twist of looking at the map from the Soviet viewpoint pulls the reader into the picture. It was used in a newspaper at about twice the size shown here.

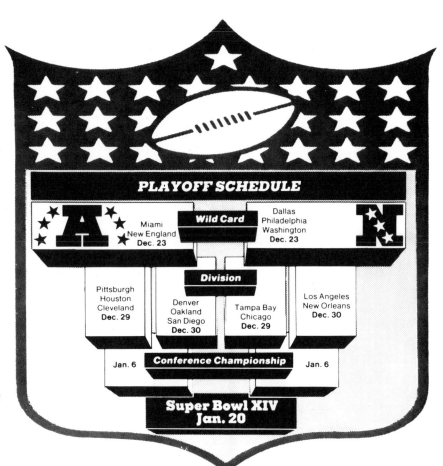

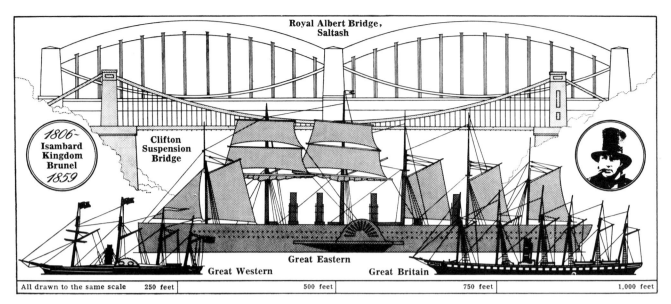

Royal Albert Bridge, Saltash

Clifton Suspension Bridge

1806~ Isambard Kingdom Brunel 1859

Great Eastern

Great Western

Great Britain

All drawn to the same scale | 250 feet | 500 feet | 750 feet | 1,000 feet

A completely visual table of comparisons, these two bridges and five ships were all the work of Isambard Brunel, and all are drawn to the same scale. The use of gradually lightening tone from foreground toward the back of the picture separates the items and gives a sense of depth.

A complete roster of the ships of the Royal Navy was drawn in silhouette in black and white for publication in England at the time of Queen Elizabeth II's Silver Jubilee. This is more a visual list than a true comparison of the ships, as it was impossible to draw them all to the correct scale. The patrol and other vessels at the bottom would have been mere pinpricks if they were to be in scale with the aircraft carrier.

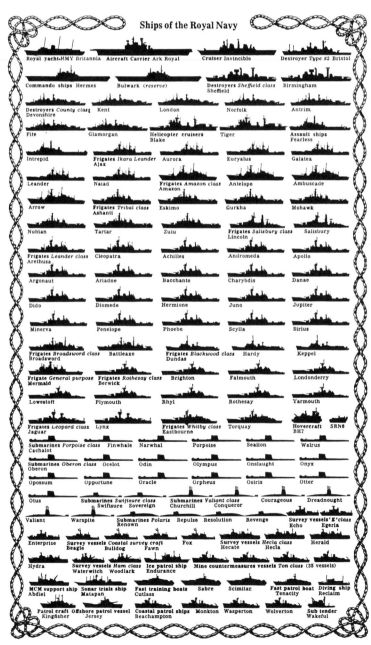

Ships of the Royal Navy

Royal yacht HMY Britannia | Aircraft Carrier Ark Royal | Cruiser Invincible | Destroyer Type 82 Bristol

Commando ships Hermes | Bulwark (reserve) | Destroyers Sheffield class Sheffield | Birmingham

Destroyers County class Devonshire | Kent | London | Norfolk | Antrim

Fife | Glamorgan | Helicopter cruisers Blake | Tiger | Assault ships Fearless

Intrepid | Frigates Ikara Leander Ajax | Aurora | Euryalus | Galatea

Leander | Naiad | Frigates Amazon class Amazon | Antelope | Ambuscade

Arrow | Frigates Tribal class Ashanti | Eskimo | Gurkha | Mohawk

Nubian | Tartar | Zulu | Frigates Salisbury class Lincoln | Salisbury

Frigates Leander class Arethusa | Cleopatra | Achilles | Andromeda | Apollo

Argonaut | Ariadne | Bacchante | Charybdis | Danae

Dido | Diomede | Hermione | Juno | Jupiter

Minerva | Penelope | Phoebe | Scylla | Sirius

Frigates Broadsword class Broadsword | Battleaxe | Frigates Blackwood class Dundas | Hardy | Keppel

Frigate General purpose Mermaid | Frigates Rothesay class Berwick | Brighton | Falmouth | Londonderry

Lowestoft | Plymouth | Rhyl | Rothesay | Yarmouth

Frigates Leopard class Jaguar | Lynx | Frigates Whitby class Eastbourne | Torquay | Hovercraft BH7 SRN6

Submarines Porpoise class Cachalot | Finwhale | Narwhal | Porpoise | Sealion | Walrus

Submarines Oberon class Oberon | Ocelot | Odin | Olympus | Onslaught | Onyx

Opossum | Opportune | Oracle | Orpheus | Osiris | Otter

Otus | Submarines Swiftsure class Swiftsure | Sovereign | Submarines Valiant class Churchill | Conqueror | Courageous | Dreadnought

Valiant | Warspite | Submarines Polaris Renown | Repulse | Resolution | Revenge | Survey vessels 'E' class Echo Egeria

Enterprise | Survey vessels Beagle | Coastal survey craft Bulldog | Fawn | Fox | Survey vessels Hecla class Hecate Hecla | Herald

Hydra | Survey vessels Ham class Waterwitch Woodlark | Ice patrol ship Endurance | Mine countermeasures vessels Ton class (35 vessels)

MCM support ship Abdiel | Sonar trials ship Matapan | Fast training boats Cutlass | Sabre | Scimitar | Fast patrol boat Tenacity | Diving ship Reclaim

Patrol craft Kingfisher | Offshore patrol vessel Jersey | Coastal patrol ships Beachampton | Monkton | Wasperton | Wolverton | Sub tender Wakeful

162

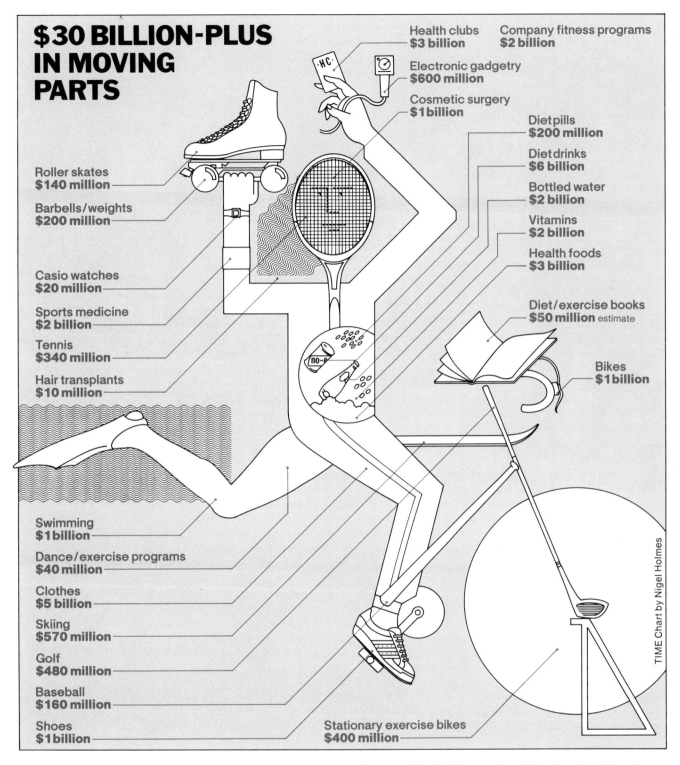

$30 BILLION-PLUS IN MOVING PARTS

Health clubs
$3 billion

Company fitness programs
$2 billion

Electronic gadgetry
$600 million

Cosmetic surgery
$1 billion

Diet pills
$200 million

Diet drinks
$6 billion

Bottled water
$2 billion

Vitamins
$2 billion

Health foods
$3 billion

Roller skates
$140 million

Barbells/weights
$200 million

Casio watches
$20 million

Sports medicine
$2 billion

Tennis
$340 million

Hair transplants
$10 million

Diet/exercise books
$50 million estimate

Bikes
$1 billion

Swimming
$1 billion

Dance/exercise programs
$40 million

Clothes
$5 billion

Skiing
$570 million

Golf
$480 million

Baseball
$160 million

Shoes
$1 billion

Stationary exercise bikes
$400 million

TIME Chart by Nigel Holmes

To accompany a photographically illustrated article about health and exercise, this stylized human figure was made out of all the financial aspects of the fitness craze. Note how the stylization of the "athlete" is carried through to the arrangement of pointing lines that are organized in such a way as to fall parallel to one another and to the lines of the figure as well, whenever possible.

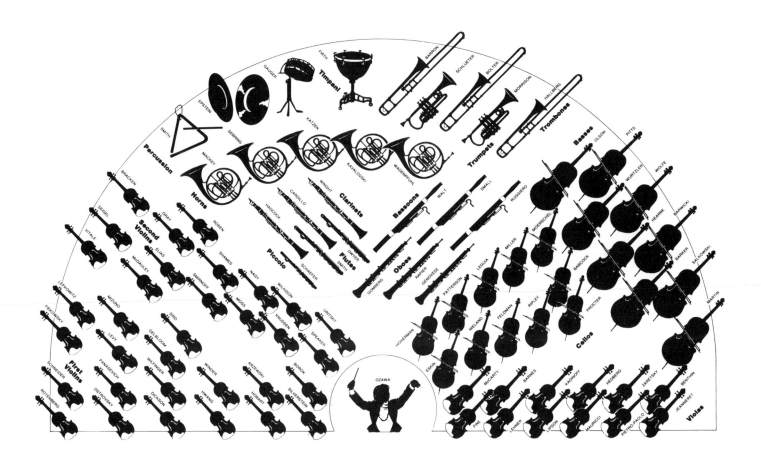

This table of contents of an orchestra has the same fascination of completeness as the Royal Navy table on page 162. The lively arrangement of beautifully simplified instruments keeps the reader interested right through to the last of the double basses!

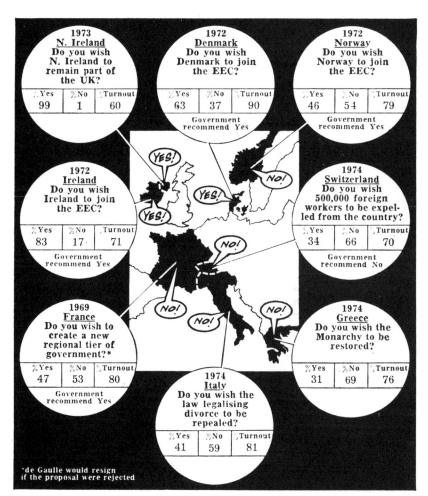

The results of a poll are tabulated here in an interesting way that shows the countries involved and cartoonlike speech balloons with undetailed "yes" or "no" answers to the questions.

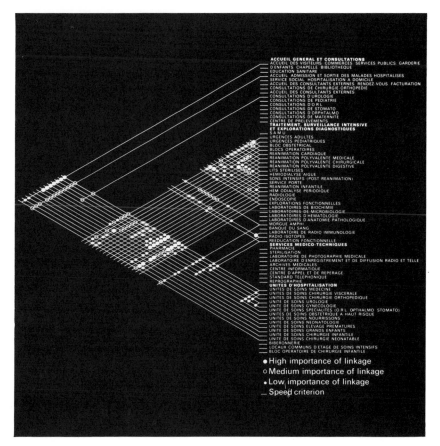

A table of intricate interrelationships is executed here with a fine sense of detail and drama. By being on a black background a graphic presence is imparted to the information. The whole piece would have been much less interesting printed in black on a white background.

Uses and Misuses

"THERE ARE THREE KINDS of lies: lies, damned lies and statistics." So said Benjamin Disraeli, and certainly his famous phrase can be borne out by examples to be found almost daily in newspapers and other publications.

It is very important for a chartmaker to check information carefully; in the rush of last-minute preparations it is easy to transpose type from, say, one bar to another. There are so many little bits and pieces on a chart and each must be in the right place for it to make sense. But more important—and this is what Disraeli meant—the artist must be responsible when it comes to designing and not deliberately try to mislead.

Too often, in an effort to dress up a chart with illustration, an artist loses the focus of the work, and the resulting art obscures the meaning of the figures, rather than helping a reader understand them. The chart may have engaged the eye, but if the information is not accurately portrayed, the artist has wasted his or her time as well as the reader's. In the long run charts will be ignored if they cannot be trusted. And they will be mistrusted if too many of them are biased, or erroneous, or too exaggerated.

There are cases when enlarging or changing scales on an adjacent chart can be justified on the grounds that a close-up detail presents the information more clearly than showing the entire scale. This will be discussed later, but the misuse of scale is probably the most common error made in preparing charts.

In this chapter, the changing of scales, among other common errors, will be examined first and then the use of deliberate distortion, whether it is justified or deceptive.

COMMON ERRORS

Errors abound in chartmaking, and nine examples will be discussed in this section.

1. Wrong Use of Scale. The scale along either the horizontal or the vertical axis can be misrepresented.

The Time or Horizontal Scale (x Axis). The periods marked along the horizontal scale must be even. That is, each one must denote the same amount of time. In (A) the line shows an even rate of increase on a chart with an uneven time scale. In (B), where the time is marked evenly, the same data show quite a different climb to the final figure. While both lines show that the figures go up, the correctly plotted version (B) gives a much more honest visual representation of the rate of increase.

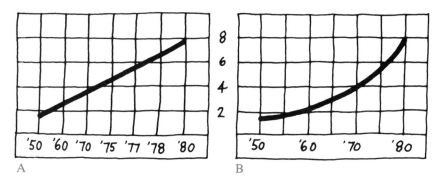

The Vertical Scale (y Axis). Changing scale on adjacent charts can give a false impression (A). If a real comparison is to be made between the two sets of statistics, they must be plotted on the same scale (B), regardless of the fact that this may seem to leave a large empty area on the chart.

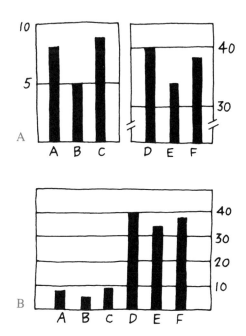

2. Wrong Choice of Chart Type. Two errors are commonly made in selecting the type of chart to be used.

Not Always a Pie Chart. Pie charts are frequently used when another form would serve the information better. As a division of a whole into its percentages the pie chart (A) is hopelessly confused and crowded in the smaller areas. The tiny difference in the smaller amounts is such that trying to separate them visually does not work. There are two possible solutions here. (B) shows that, if permitted, one way of solving the crowding problem is to lump together in one larger category all those quantities below a certain amount, in this case 2.

However, if the smaller amounts are considered too important to be lost in an "other" category, the solution lies in changing the form of the presentation to a table (C).

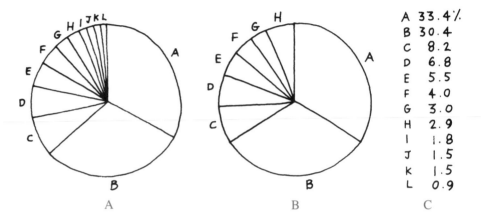

A	33.4%
B	30.4
C	8.2
D	6.8
E	5.5
F	4.0
G	3.0
H	2.9
I	1.8
J	1.5
K	1.5
L	0.9

    A   B   C

Form Versus Grid. Confusion arises when the form used conflicts with the grid. Instead of plotting the information as thin bars (A), replot it as a fever line (B) or restrict the grid (C). See the following discussion of grids.

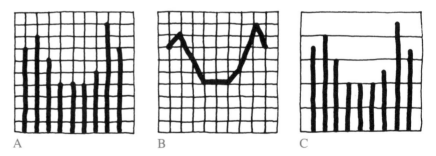

    A   B   C

3. Grid. The question here is judging what is needed to make the grid work. Often the answer doesn't really apply.

Too Many Lines. When the entire structure of the underlying grid is printed it disturbs the eye so much that the information is obscured (A). Leave out selected parts of the grid (B).

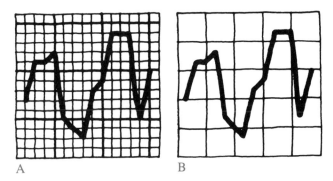

Not Enough Grid Lines. There must be enough reference points to help the eye move to any part of the chart quickly. In (A) it is difficult to read the quantities as the line floats in an empty space. This error is particularly evident when a fever line is used, but less so with a bar chart (see the second example under 2 above). For fever charts put in some of the grid (B).

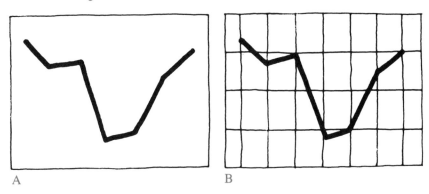

Decorative Grids. In (A) the grid has no relation to the information carried by the bars. It is purely decorative—to give the chart more of a "chartlike" look. (B) shows the grid correctly drawn.

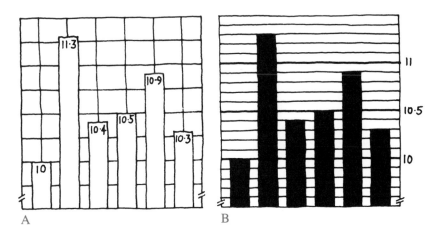

4. Typeface Considerations. Judging the spacing and size of the type can often go wrong.

Too Much Space between Columns. The eye has to travel too far in (A) to connect the subject with its quantity. (B) shows a better version. You also have the option of using the table without the illustration on the sides, thus saving space, if necessary.

Type Too Small. While the type in (A) may have been readable when it was prepared, after reduction to the printed size it clearly is not. Leave out some of the numbers to allow space for those remaining to be of sufficient size to be read (B).

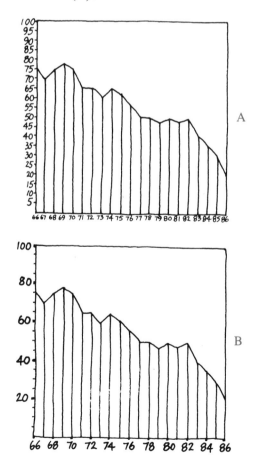

170

5. Shading. This factor, often overlooked, can dilute the effect you want or make one you don't intend.

Not Enough Contrast. In (A) the tonal values of the different parts are not sufficiently separate. Make the contrast more visible (B).

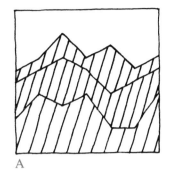 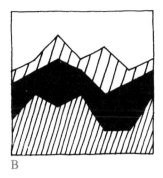

Optical Illusions. The diagonal shading in (A) gives the bars a crazy drunken look. (Remember that when you're doing a chart about the rise in alcoholism!) If it's necessary to differentiate the bars make sure the contrast is achieved by avoiding a mixture of diagonals (B). If it is not necessary to differentiate the bars, shade them all at the same angle and the bars will remain sober (C).

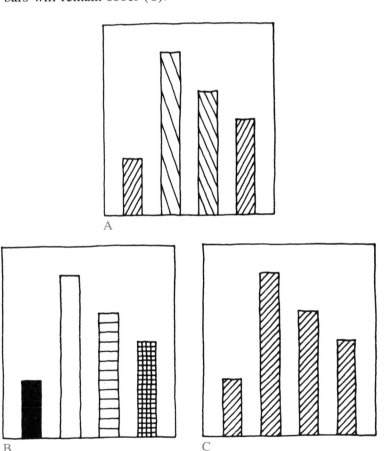

6. Imprecise Point of Reference. A mistake here can lead to serious misinterpretations.

Lines Used Are Too Thick. In an effort to be bold and eye-catching, the thickness of the line in (A) has obscured the figures it represents. A thinner line is better (B).

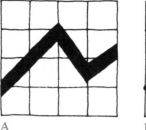 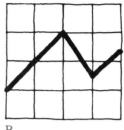

A   B

Shaped Bars. Rounded bars give the wrong impression of the information, because only a tiny part actually reaches the height that the whole bar is supposed to reach (A). Use a bar with a flat top or something with a definite "reading point" (B).

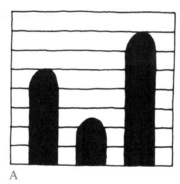 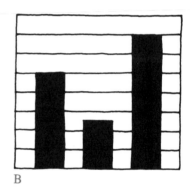

A   B

Three-Dimensional Problems. Be clear about which part of the drawing is the part that records the information. In (A) there are three possible choices of reading point. (B) shows a possible solution: Filling in the shape makes the whole thing virtually solid, with less prominence given to the line of the front.

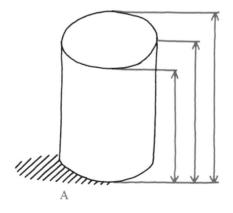 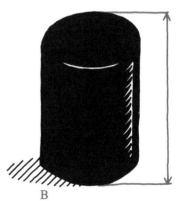

A   B

7. Labels. Three things often go wrong in labeling artwork.

Inconsistency. By leaving the type out of only one of the bars, too much prominence is given to it, since it is the odd one out (A). Rearrange the type so that all the labels are in a consistent relationship to their bars (B).

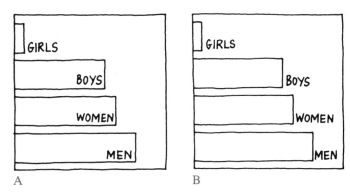

Headings. Do not assume that the reader can distinguish between two scales, even with a headline (A). Label both lines (B); they are not redundant.

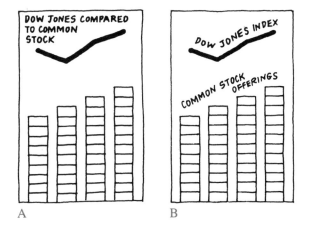

Position of Scale. On the basis that most people want to read the latest information on the given subject, put the scale on the outside right-hand edge (B), not on the left (A).

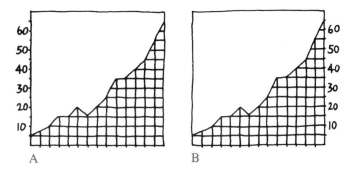

8. The Non-Bar Chart. When a bar is inserted between a name and a quantity the result can be surprisingly misleading (A), as the immediate visual impression can be exactly the opposite of what is intended. Do not use bars decoratively. Rearrange the information, for example, by hanging them off a central line (B) and rank them by quantity or by alphabetical order as shown here.

A   B

9. Proportion. Enlarging the drawing in (A) so that its height is twice as great should never be used to denote that the quantity has doubled. Since the other dimensions of the drawing have also increased, the area now shown is actually four times as large! A better way to show the increase pictorially is to use the drawing as a unit and to repeat the unit to the correct height (B).

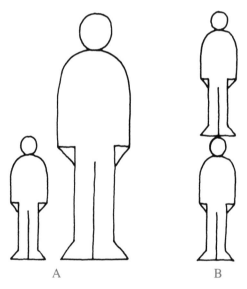

A   B

In the case of the two circles below, it can of course be argued that although it is strictly true mathematically, the large circle does not look twice the size of the small one. Points of view differ about the allowability of a certain amount of exaggeration for visual effect in such a case. The trouble is that one person's view of exactly how much bigger the larger circle can become so that it does look twice as big as the small may not coincide with another person's idea of that size.

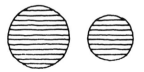

If clearly labeled, some flexibility should be permitted for an artist to use what is after all a very dramatic visual comparison technique.

DISTORTION

Here are two quick examples of distortion.

■ In a two-car race between the U.S.A. and the USSR, the American wins. According to the Russian report the Russian car finished just behind the leader, and the American car came second to last.

■ By the law of averages, the man with one foot in boiling water and the other in freezing water is pretty comfortable.

In the first propaganda has clearly got the better of the chauvanist reporter. In the second too literal an adherence to a law or system of science has produced a peculiarly painful result. Both examples show how distortion can change a set of statistics. As was stated in Chapter 3, distortion is unacceptable when it does in fact *distort* the data.

Deliberate Scale Change. If you need to illuminate a point by coming in closer to the information, then do it. Purists will say this is distortion. It is. You are deliberately distorting to make the facts clear. The facts are the same in (A) and (B), but they are visibly clearer in (B).

Where the latest figures in a sequence are the most important, but where an historical review before them is also useful for context, a case can be made for a gentle type of distortion by perspective.

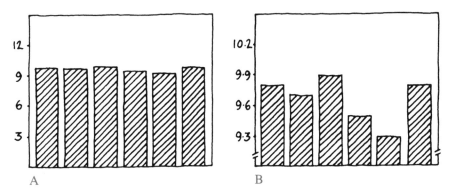

The convention of perspective is so much a part of what we see and experience that the eye can take in all the information easily, so it is possible to use less space by showing the first part of the sequence in perspective.

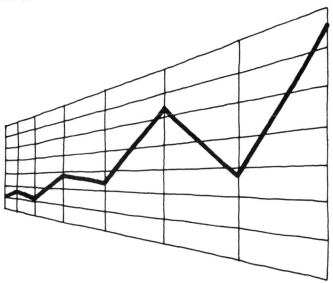

Deceptive Distortion. An example of this is drawing bars of different widths. In (A) a selected bar has been doubled in width to make it stand out. This is a very bad practice. In (B) the highest bar is still the highest, but now it is fairly shown to be so. It is graphically acceptable to color this bar differently for emphasis, where, for instance, one product's sales are to be contrasted with another's.

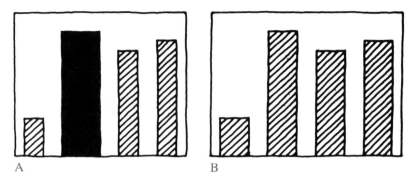

Here there is too much dependence on nonvisual means to convey the whole message (A). The tiny asterisk on the third column notes that the time span of the figures for that bar are for half a year, compared with the others that are for a whole year. The overriding visual effect is that the bars are to be read as you see them. It is wrong to assume that an asterisk can correct the discrepancy in the visual presentation by meekly stating the fact underneath it. Charting is a visual presentation; too many footnotes defeat the purpose. It is more honest to project the rest of the year with a dotted line (B), thus showing a completely different picture.

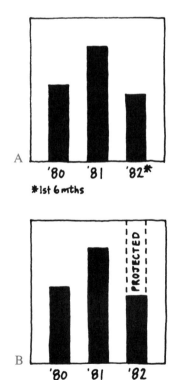

By manipulating the scale, figures can be made to do things that they plainly should not. Consider this table of figures:

Year	Percentage	Year	Percentage
1975	5	1979	5.7
1976	5.1	1980	5.8
1977	5.2	1981	5.9
1978	5.4	1982	5.95

There are two ways of charting them. The deception, however, really comes in the titles. (A) might be titled "massive increase in unemployment." (B) might be titled "unemployment stays level."

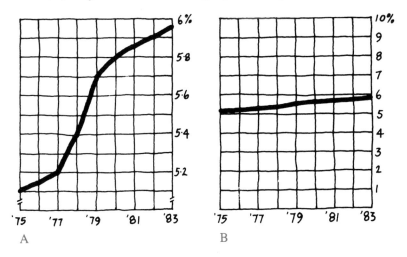

A   B

When used as propaganda, statistics can be dangerous. Darrell Huff's book *How to Lie with Statistics* will show you most amusingly what can be done.

Chapter Seven
Found Charts

BY NOW I HOPE that you will be sufficiently in tune with my thinking about charts to forgive a little indulgence in what may be termed "found" examples of statistical presentation.

In everyday life there are changes in nature: as things grow they leave evidence of their past. For example, veterinarians can tell the age of a horse from the size of its teeth—the horse's mouth is a chart of a kind. Just how "long in the tooth" is a measurable statistic. Doubtless doctors could tell the ages of their patients the same way. Of course when they are alive it's much easier to ask them, but it is a valuable skill for a coroner. As for lying about your age . . . just don't do it to a vet!

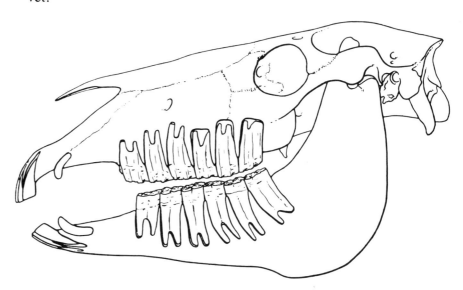

As a horse ages, so its teeth change—the surfaces are worn down and the roots grow longer.

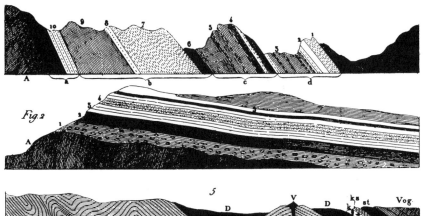

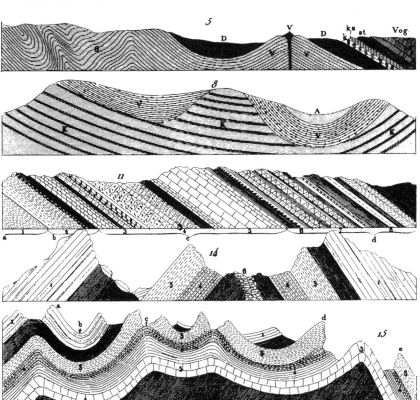

In certain geological formations nature has left graphic details of what it was like in ages past. The familiar geological patterns that are known from school textbooks are perfect charts of the earth's history. The geologist does not even have to translate abstract data into a visual form; he or she merely mimics nature itself in diagrams of layers of the earth's crust.

If the slices through the earth can be claimed to be the world's largest charts, perhaps a different claim can be made for another of nature's chart efforts—the inside of a tree. This is surely the most successful graphic presentation of age and growth. And with considerable subtlety and detail too. Apart from counting rings to show age in years, one can also learn when there were lush, fertile, rain-filled years (wider rings, equaling more growth) and when there was a drought, restricting the growth and leaving a thinner more closely spaced ring.

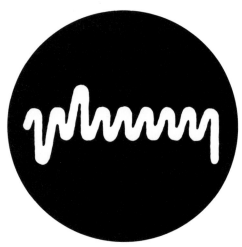

Sometimes a graphic form may be used for its own sake. The electronics company Plessey has a logo that is itself a graph line spelling out the name.

2-DRAWER
LETTER LEGAL

SALE	SALE
$94	$99

4-DRAWER
LETTER LEGAL

SALE	SALE
$119	$139

5-DRAWER
LETTER LEGAL

SALE	SALE
$165	$189

Other found examples simply look like charts: This advertisement for filing cabinets ought to have some information in it!

Modern suspension bridges display most honestly the engineer's science of stress related to holding up a roadway over a long span.

The skyline of many cities can represent, in the form of a bar chart, a number of different statistics. For example, the height of the buildings in the Manhattan skyline is directly related to (1) the amount of people working from 9 to 5 on Monday to Friday; (2) the amount of money that is being discussed or dealt with daily; (3) the value of real estate in the area; (4) the consumption of martinis (in multiples of three, of course) at lunchtimes; (5) the amount of energy being consumed by computers, word processors, air conditioners, electric typewriters; (6) the accumulated decibels of phones ringing; or (7) the frustrations of people on the street who cannot see the sun.

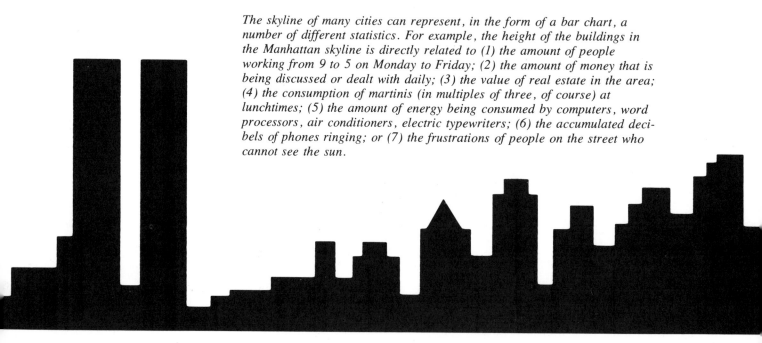

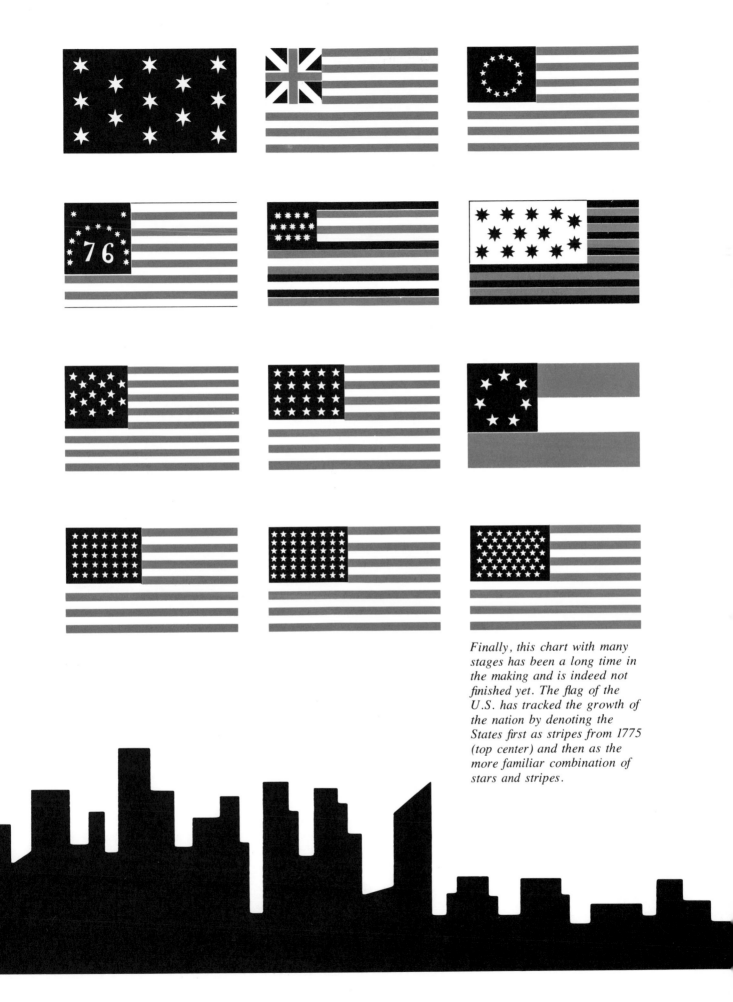

Finally, this chart with many stages has been a long time in the making and is indeed not finished yet. The flag of the U.S. has tracked the growth of the nation by denoting the States first as stripes from 1775 (top center) and then as the more familiar combination of stars and stripes.

Glossary

Terms in italics are defined herein.

Abscissa. *Horizontal* or *x axis* of *graph*. One of the *Cartesian coordinates*. Generally the baseline of a graph and the axis most commonly used for marking off time periods.

Acetate. Sheet of thin transparent film. Clear acetate is used for protecting or covering artwork. Prepared or frosted acetate is used for drawing; it is still transparent, though less so than the clear acetate. The slightly cloudy appearance of the prepared side is really a roughening of the surface, which allow the pens used by draftsmen and artists to "bite" into it.

Amberlith. Tradename for material used in preparation of overlays for *preseparated artwork*. A thin sheet of *acetate* to which is applied a layer of amber-colored film. This layer can be cut, and the unwanted parts of the film stripped away, without harming the carrier acetate. The amber color is photographically opaque, while remaining transparent to the eye. It is also made in a darker red called "Rubylith."

Artwork. The work that the chartmaker (or any commercial artist or illustrator) prepares for the printer. The finished chart (drawing, type, and *overlays* mounted together on a board) are "marked up" with printing instructions. The entire package is termed the "artwork."

Average, Mean, Median, Mode. A value that represents a group of others.
1. The arithmetic average or mean is found by adding together all the numbers in a series and dividing by the amount of numbers. For example, in this series:

$$12222222344557789$$

the average is 3.94, which is reached by adding all of them together (67) and dividing them by the amount of numbers in the series (17). Thus $67 \div 17 = 3.94$.
2. The median number in the same series is 3. This is the number in the middle of the series, with eight numbers before it and eight numbers after it.
3. The mode number in the same series is 2. This is the number that occurs most often in the series.

Bar Chart. Graphic presentation of statistics in the form of abstract bars or columns. A pictorial bar chart substitutes drawings of the objects being charted for the abstract bars. In both cases the height of the bar/drawing represents the magnitude of the statistics.

Bear Market. Stock market phrase to describe a continuing trend of falling stock prices. (Remember it by "bear down.")

Bull Market. The opposite of a bear market, this describes a strong market, where an index of performance (for example, the Dow Jones Industrial Index) is rising.

Camera-Ready Art. *Artwork* completely prepared for the platemaker's camera; no additional sizing or stripping in of other images is needed.

Cartesian Coordinates. Named after Rene Descartes (1596–1650), French philosopher and mathematician. The two lines that are the basis of all graphing. Drawn at right angles to one another, the vertical line usually represents the quantity of the substance being plotted, while the horizontal represents the time frame of the chart. See *vertical (y axis/ordinate)* and *horizontal (x axis/abscissa)*.

Continuous Tone Original. Photograph or drawing that has no *halftone* screen on it.

Curve. The line that joins together the plotted points of a *fever chart*. Thus it is synonymous with fever line or graph line.

Divided Circle. More commonly called a "pie" chart. It takes a whole circle to represent 100 percent, and the segments of it represent divisions of that 100 percent.

Drafting Film. Non-stretching or -shrinking *acetate* prepared on one or both sides with a surface that allows an artist to draw on it while still seeing through the film to a drawing below it.

Fever Chart. Chart prepared on graph paper by plotting points from two *coordinates* generally representing time and quantity and then joining up those points to form one continuous line or *curve*.

Final Art, Finished Art. See *Artwork*; they mean the same thing.

Flowchart. A sequence of events arranged graphically in the order in which they take place. If a time scale is included, part of the flowchart may be highlighted as the "critical path," which shows the most important steps that must be followed for the job to be completed properly and/or on time.

French Curve. Plastic *templates* with a variety of swinging curves that enable the artist to draw smooth, nonfreehand, continuously sinewy lines.

Graph. Visual representation of two sets of figures. Usually time and quantity applied to the price of a commodity, sales of a product, or changes in a statistic.

Grid. The basis, and often background, for a *graph*. The grid lines correspond to the time period and quantity of the matter being charted. When shown in the final printed image, they help the eye join up the scales on the *coordinates* to the plotted points.

Halftone.
1. The graphic design and printer's term for a photograph, which includes a tonal range from white or any degree of gray to black.
2. The photograph when it is actually printed by means of a halftone screen. This breaks up the tones of the original into tiny dots of varying size, which are not apparent to the naked eye.

Horizontal Scale. Base line of chart or *graph*, also called the *x-axis* or *abscissa*, on which are generally marked divisions of time.

Key or Legend. Used to differentiate between colors or types of line and positioned on the chart in an unobtrusive place. A description of what the line represents is written out next to a small sample of the color or type of line.

Labels. The notations that inform the reader about the subject matter and details of the *graph* or chart.

Legend. See *Key*.

Line Art. That which can be reproduced in "line," that is, with no "tonal" values at all (just solid black and white; no grays).

Line Chart. The same as *fever chart* or line *graph*. A single line joining together a number of points plotted on a *grid* to show a trend.

Logarithmic Scale. A scale that allows for (1) plotting ratio charts and (2) plotting a series of numbers, some of which are very small and some very large. The appearance of logarithmic paper is unlike normal graph paper in that the horizontal lines are not equally spaced.

Matrix. The basis for most tables. A *grid* of spaces that can be filled in with information.

Multiple Bar Chart. *Bar chart* with layers of information arranged on each bar, showing the breakdown, or component parts, of a substance for the same time period.

Multiple Fever Chart. More than one fever line plotted on the same *grid*: the activity of more than one set of statistics over the same time period.

Mylar. Tradename for prepared or frosted *acetate*. Available in different weights and in a range of sheet sizes or in a continuous roll.

One-Time Method. A term that describes a piece of work that is not going to be printed or published and therefore exists only as itself. A hand-drawn chart; a chart prepared with graphics aids for a presentation.

Ordinate. The *vertical* or *y axis* of a *graph*.

Organization Chart. An arrangement of boxes connected by lines. Each box usually contains a name or department title, and the connecting lines show their hierarchical relationship.

Overlay. Those parts of the *artwork* that indicate the colored areas. Can be cut from *Amberlith* (for larger unmodulated areas) or drawn directly on drafting film (for fine lines or lettering). Overlays are aligned with the base artwork or drawing and with other overlays by use of *register marks*.

Peaks. High points of a *fever chart*.

Pica. Printers' unit of measurement. One pica equals about $1/6$ of an inch and is divided into 12 points. Type sizes are in points, never picas, even when they exceed 12 points. Thus 36-point type is never referred to as 3-pica type.

Pictograph. A chart or *graph* using symbols or symbolic pictures instead of abstract bars.

Pie Chart. See *Divided Circle*.

Plot Point. The intersection of two pieces of information plotted on a *grid* or graph paper. When a number of such points are joined up a *fever chart* is produced.

Preseparated Art. *Artwork* that the printer does not have to separate photographically in order to print it in color. The areas in a drawing or chart that will be colored in the final printed image are prepared by the artist on separate *Amberlith overlays* and mounted on top of a base drawing. The opposite of preseparated art is called "reflective art." This has all the colors and type painted on one surface. Although as a piece of artwork it bears more resemblance to the finished printed image, and thus is useful to a client in his or her understanding of how the job should turn out when printed, it invariably produces a less crisp end result.

Pressure-Sensitive Graphics.
1. Letters and numbers lightly adhered to a clear plastic carrier film, which can be burnished down onto almost any flat, smooth surface.

2. Sheets of tone, either continuous or broken down into specified screens, which can be burnished down lightly onto a drawing, cut to the shape required, with the excess peeled off and the remainder pressed down hard.

3. Charting tapes made in a variety of widths and patterns and dispensed from a roll. They are cut to the length needed and burnished down onto the surface of the art. There are many companies producing pressure-sensitive graphics, among them Artype, Cello-Tak, Chartpak, Copyaid, Formatt, Letraset, Mecanorma, Prestype, Tactype, Zipatone.

Process. Method by which a printer separates reflective or other original art that is not already preseparated. The process camera divides the colors of the original into percentages of yellow, red, blue, and black, which are called yellow, magenta, cyan, and key. A separate printing plate is made for each color.

Projection. An estimated or projected number at the end of the chart to show the continuing trend after the last known figure. Should always be made graphically different from the rest of the chart, for instance, by dotting the line in a *fever chart* or shading the bar in a *bar chart*, and by labeling it "projected" or "estimated" in both cases.

Register/Registration Marks. Marks applied to *overlays* to make sure that they align with one another and with the base *artwork* or drawing. They can be drawn onto the *overlays* (a simple cross) or applied from a sheet of Letraset or similar pressure-sensitive material. The best and easiest way, however, is to buy them preprinted on a roll of tape.

Source. Important but often neglected part of the labeling of a chart. Tells the reader where the information came from that enabled the chart to be made. So as not to be confused with the credit line for the piece, it is generally written: "source: (name of source)."

Technical Pens. Delicately engineered drawing pens with various line thicknesses, which start at .13 of a millimeter and go up to 2 millimeters. The drawn lines are consistently even. Several makes are available; it is largely a matter of personal taste as to which to use. Compass attachments are available for drawing circles of the same line width as the rest of the art.

Template. Clear, tinted plastic shapes used with technical pens to draw perfect circles, ovals, *curves*, and lettering.

Title. Important label to introduce the reader to the subject. A well-written title should sum up the intent of the chart.

Triangle. Right-angled piece of plastic or aluminum, with either two 45-degree and one 90-degree angle or 30-, 60-, and 90-degree angles. Sometimes beveled for easier inking. Adjustable triangles are extremely useful where an angle other than 30, 45, or 90 degrees is needed.

Troughs. Low points of a *fever chart*.

Vertical Scale. Used on a *graph* to plot the quantity of the subject. The *y axis* or *ordinate*.

X-Axis. The *horizontal axis*, or *abscissa*. One of the *Cartesian coordinates*.

Y-Axis. The *vertical axis*, or *ordinate*. One of the *Cartesian coordinates*.

Selected Bibliography

Brinton, Willard C. *Graphic Methods for Presenting Facts*. New York: The Engineering Magazine Company, 1914.

Enrick, Norbert Lloyd. *Effective Graphic Communication*. Pennsauken, NJ: Auerbach Publishing, 1972.

Hart, Ivor B. *The Mechanical Investigations of Leonardo da Vinci*. Berkeley and Los Angeles: University of California Press, 1963.

Herdeg, Walter. *Graphis Diagrams*. Zurich: Graphis Press, 1981.

Huff, Darrell. *How to Lie with Statistics*. New York: W.W. Norton & Co., Inc., 1954.

Monkhouse, F.J., and H.R. Wilkinson. *Maps and Diagrams: Their Compilation and Construction*. 3rd ed. London: Methuen & Co. Ltd., 1971.

Peddle, John B. *The Construction of Graphical Charts*. New York: McGraw-Hill Book Company Inc., 1910.

Playfair, William. *The Commercial and Political Atlas*. London: Debrett, 1787. Additional charts by James Corry.
————.*The Statistical Breviary*. London: T. Bensley, 1801.
————.*An Inquiry into the Permanent Causes of the Decline and Fall of Powerful and Wealthy Nations*. London: 1805.
————.*The Translation of Dennis Donnant's "Statistical Account of the USA."* London: Greendland & Norris, 1805.

Raisz, Erwin J. *General Cartography*, 2nd ed. New York: McGraw-Hill Book Company Inc., 1943.

Selby, Peter H. *Interpreting Graphs and Tables*. New York: John Wiley & Sons, 1976.

Tschichold, Jan. "Statistics in Pictures," in Vol. 11, *Commercial Art*. London: The Studio Ltd., 1931.

Tufte, Edward. *The Visual Display of Quantitative Information*. Cheshire, CT: Graphics Press, 1983.

Weld, Walter E. *Principles of Charting*. New York: Barron's, The National Financial Weekly, 1939.

Zelazny, Gene. *Choosing and Using Charts*. New York: McKinsey and Co., 1972.

SELECTED READINGS ON SYMBOLS

Dover Pictorial Archive Book Catalog. New York: Dover Publications.

Dreyfuss, Henry. *Symbol Sourcebook*. New York: McGraw-Hill, 1972.

Modley, Rudolf. *Handbook of Pictorial Symbols*. New York: Dover Publications, 1976.

Random House Encyclopedia.

Graphic art catalogs: Artype, Chartpak, Copyaid, Formatt, Letraset, Mecanorma, Zipatone.

Other mail-order catalogs: Brookstone, L.L. Bean, Sears.

Credits

Unless otherwise noted, all charts and other illustrations are by Nigel Holmes.

The following charts by Nigel Holmes were originally published in *Time* magazine: page 29 (top), 30 (top), 31 (bottom), 32, 34, 38 (bottom), 41 (top right), 42 (both), 43, 44, 45, 47 (bottom), 51, 52 (bottom), 55 (right), 56 (top), 57 (both), 59, 64 (left), 68, 69, 77, 78, 80, 87, 92, 105, 115, 116 (both), 121 (bottom), 122 (left), 123 (top), 124 (both), 125 (both), 129 (bottom), 132, 138, 139, 140, 141 (both), 143, 146, 147 (both), 148 (top), 150 (bottom), 151 (both), 154, 155 (top), 156, 158 (both), 163.

12 (top) Leonardo da Vinci, *Notebooks*, folio 44 verso.
15 (top) William Playfair, *An Inquiry into the Causes of the Decline and Fall of Powerful Nations*, 1805.
15 (bottom) William Playfair, *Commercial and Political Atlas*, 1787.
16 (top) William Playfair, *Commercial and Political Atlas*, 1787.
16 (bottom) William Playfair, *Statistical Account of the United States of America*, 1805.
17 (both) William Playfair, *Statistical Breviary*, 1801.
18, 19 Irving Geis, *Fortune* magazine.
20 (both) Otto Neurath, Vienna Museum of Social and Economic Studies.
21 (top) *Our City, New York*. Published by Allyn & Bacon.
21 (bottom) *Der Spiegel*.
28 Richard Yeend, *The New York Times*.
30 (bottom) Esso.
31 (top) Basic Four Corporation.
33 (top) *USA Today*.
38 (top) Prince.
39 (top) Gips & Balkind Inc.
39 (bottom) Harrah's.
40 (top) *USA Today*.
40 (bottom) Richard Leadbetter, *The Times*, London.
41 (top left, bottom) Japan Statistical Association.
46 (top) Scott McNeil, *The New York Times*.
46 (bottom) Robin Jefferies, *Kansas City Star*.
47 (top) *Panorama* magazine.
48 (bottom left) Moore McCormack Resources, Inc.
48 (bottom right) *Folio* magazine.
49 (top) Diagram Visual Information.
49 (bottom) William Reduto, *Fortune* magazine.
50 (top) John Grimwade, *The Times*, London.
50 (middle, bottom) Richard Yeend, *The New York Times*.
52 (top left) *Places Rated Almanac*.
52 (top right) American Forest Institute.
53 Richard Yeend, *The New York Times*.
54 Programma Lineor.
55 (top) Club Mediteranée.
58 British Tourist Authority.
62 Letraset.
63 (top) SAS graph.

Nigel Holmes is Graphics Director of Time Magazine. A graduate of the Royal College of Art in London, with a Master's Degree in Illustration, he has been with Time since 1978. He does freelance design for magazines and for advertising campaigns of several major U.S. corporations including IBM, Texaco, Chase Manhattan, and Merrill Lynch.

Since 1984 Holmes has run an annual workshop for professional information designers at the Rhode Island School of Design, and he teaches at seminars around the United States.

Index

Page numbers in italics refer to illustrations.